MATH SHORTCUTS

FLORA M. LOCKE

Merritt College
Oakland, California

John Wiley & Sons, Inc.

New York • Chichester • Brisbane • Toronto • Singapore

Library of Congress Cataloging in Publication Data

Locke, Flora M
Math shortcuts.

(Wiley self-teaching guides)
1. Ready-reckoners. I. Title
QA111.L7 513'.92 72-5775
ISBN 0-471-54328-4

Printed in the United States of America

72 73 10 9 8 7 6

Preface

In this book you will learn many shortcuts to basic mathematics--methods that will increase your speed and accuracy in using numbers and your ability to analyze and solve word problems. Adding, subtracting, multiplying, and dividing are all made easier. Along the way you will review some of the "problem" areas in basic mathematics such as fractions and percentages. Also, you will learn through estimating answers to avoid serious errors in your calculations.

Working with numbers when you use shortcuts can be both fun and highly profitable in school or on the job. Employers and teachers are quick to notice facility with numbers. You can also use these shortcuts in your personal finance and everyday calculations. Finally, using shortcuts will help you develop a greater appreciation of number relationships, an appreciation that will increase self-satisfaction in your work.

This book began as the first two chapters of BUSINESS MATHEMATICS (also a Wiley Self-Teaching Guide). Students found it so useful that I reorganized and expanded the material for those who want to learn shortcuts to solve everyday math problems.

Sebastopol, California
August, 1972

Flora M. Locke

To the Reader

Your use of this book will depend, of course, on your facility with numbers and your knowledge of number relationships. Objectives are listed at the beginning of each chapter that describe what you will be learning. Many pretests are offered; the results of these pretests will indicate what sections you need to study. If you do well on a pretest, you may bypass some of the material.

In a programmed book, you actively participate in learning the material. Each chapter is arranged in units that are broken down into short steps called frames. After each explanation and example, you are asked a question or given a problem to solve. After you have written your answer in the space provided, compare it with the one given below the dashed line. If the answers do not agree, read the frame again and check your arithmetic. If they still do not agree, review the preceding frames. For more effective learning, it is best to cover the answers (below the dashed line) with an index card or a folded piece of paper while you are working the problems.

At the end of each chapter you will find a test for self-evaluation. Answers are provided for all problems, and solutions are also offered where appropriate. Check your arithmetic if your answer to a problem does not agree with the one given. If the answers still do not agree, refer to the solution. Frame references are included with the answers for these tests in case you wish to review certain sections.

The last chapter, "Solving Word Problems," puts to use the lessons you learned in the preceding chapters. You will find answers as well as complete solutions to all of these problems.

The Appendix offers additional exercises if you want more practice.

F. M. L.

Contents

CHAPTER ONE

Addition and Subtraction (Whole Numbers and Decimals)

OBJECTIVES

There are many shortcuts you can use in adding and subtracting. In this chapter you will learn to:

- group by tens in adding;
- use subtotals in adding;
- use reverse-order checking in adding and subtracting;
- round a number to any given place;
- estimate answers in addition and subtraction.

You will have many chances to skip material you already know.

Unit 1

READING AND WRITING NUMBERS

1. First, let us see if we agree on some basic terms. Underline the correct answer for each of the following:

   25 is a (whole/decimal/mixed) number.
   31.6 is a (whole/decimal/mixed) number.
   .27 is a (whole/decimal/mixed) number.

   - - - - - - - - - - - - - - -

   whole, mixed, decimal

   If your answers are correct, skip to Frame 6. Otherwise, continue on here.

2. Our number system is called the decimal system (from the Latin decem, which means ten), because it is based on 10 symbols or digits--0, 1, 2, 3, 4, 5, 6, 7, 8, 9. The base is 10 because it takes 10 in any one place to equal 1 in the next higher place; for example, it takes 10 ones to equal 10, 10 tens to equal 100, 10 hundreds to equal 1000.

3. A number consists of one or more digits. A whole number is a digit from 0 to 9 or a combination of two or more digits such as 25 or 1037.

   231 is a ______________ number.

   - - - - - - - - - - - - - - - -

   whole

4. A decimal, which is less than a whole number, is indicated by the decimal point and may consist of one or more digits. Examples are .5, .52, .678, .008.

   .059 is a ________________ number.

   - - - - - - - - - - - - - - - -

   decimal

5. The decimal point separates whole numbers from decimal numbers. A number such as 23.056 is called a mixed number, since it consists of a whole number part (23) and a decimal part (.056).

   52 is a ______________ number.

   2.067 is a ______________ number.

   .825 is a ________________ number.

   - - - - - - - - - - - - - - - -

   whole, mixed, decimal

6. How good are you at translating numbers into words, and words into numbers? Try these to see if you can skip the next section.

   A. 3.06 is read "three and six hundredths."
      3108 is read "three thousand, one hundred eight."
      The following numbers are read as

      15 ____

      22.5 ____

306.03 ______________________________

2540.096 ______________________________

.10035 ______________________________

B. Nine and seventy thousandths written as a number is 9.070. Thirteen thousand seven hundred fourteen is written as 13,714. Write the following as numbers:

two and nine-tenths __________

three thousand fifty-six __________

eighty-two ten thousandths __________

nine thousand, four hundred ten, and one hundred twenty-two thousandths __________

two hundred ninety-eight thousand, six hundred two and two-tenths __________

A. 15 is read as "fifteen."
22.5 is read as "twenty-two and five tenths."
306.03 is read as "three hundred six and three hundredths."
2540.096 is read as "two thousand five hundred forty and ninety-six thousandths."
.10035 is read "ten thousand thirty-five hundred thousandths."

B. 2.9; 3056; .0082; 9,410.122; 298,602.2

If your answers were all correct, skip to Frame 12. If you made any errors or had any difficulty, continue on here.

7. Digit positions to billions (whole numbers) and billionths (decimal numbers) are illustrated in the chart below:

Billions	Hundred millions	Ten millions	Millions	Hundred thousands	Ten thousands	Thousands	Hundreds	Tens	Units	DECIMAL POINT	Tenths	Hundredths	Thousandths	Ten-thousandths	Hundred-thousandths	Millionths	Ten-Millionths	Hundred-millionths	Billionths

←Whole Numbers →·← Decimal Numbers→

Examples in reading:

8 is read "eight."
23 is read "twenty-three."
458 is read "four hundred fifty-eight."
6108 is read "six thousand, one hundred eight."

The word "and" is used to indicate the decimal point for numbers that include digits to the right of the decimal; for example,

2.5 is read "two and five tenths."
167.09 is read "one hundred sixty-seven and nine hundredths."
6030.4 is read "six thousand thirty and four tenths."
.305 is read "three hundred five thousandths."

8. The following numbers are read as

2.6 ________________________

30.07 ________________________

.08 ________________________

400.56 ________________________

10,906.7 ________________________

- - - - - - - - - - - - - - -

"two and six tenths," "thirty and seven hundredths," "eight hundredths," "four hundred and fifty-six hundredths," "ten thousand, nine hundred six and seven tenths"

9. Zeros may be added after the decimal point without changing its value. Thus 2.5 = 2.50 = 2.5000, etc., 56 = 56.00, etc. Also, .5 = five tenths or fifty hundredths (the value is the same).

.03 = ____________ hundredths or ______________ thousandths.

12.5 = twelve and ___________ tenths or twelve and ____________ hundredths.

.5500 = five thousand, five hundred ten-thousandths, or

______________________ hundredths.

- - - - - - - - - - - - - - -

three or thirty; five or fifty; fifty-five

10. Large numbers are generally easier to read when the comma is used to group hundreds, thousands, millions, and so on. For example, the number 3678458 is easily and quickly read when written as 3,678,458; that is, "three million, six hundred seventy-eight thousand, four hundred fifty-eight."

4589065 written as 4,589,065 is read ______________________

____

- - - - - - - - - - - - - - - -

four million, five hundred eighty-nine thousand, sixty-five

11. Eight hundred twenty-five and sixteen thousandths is written as

______________. Nine thousand fifty-six and eight ten-thousandths is written as ______________. Ten thousand, one hundred six and two hundred fifty thousandths is written as

______________.

- - - - - - - - - - - - - - - -

825.016; 9,056.0008; 10,106.250

Unit 2

ADDITION

12. To see if you can skip this next section, try these two problems. Arrange the following numbers in a column and add:

(a) 13.75, 8.9, .006, 52.009, 21, 5.6098

(b) Two and nine-tenths
Three thousand fifty-six
Eighty-two ten-thousandths
Nine thousand
Four hundred ten and one hundred twenty-one thousandths

(a)	(b)
________	
________	________
________	________
________	________
________	________
Totals ________	________

(a)
```
 13.75
  8.9
   .006
 52.009
 21.
  5.6098
--------
101.2748
```

(b)
```
     2.9
 3,056.
      .0082
 9,000.
   410.121
-----------
12,469.0292
```

If either answer was incorrect or if you had any difficulty, continue on here. Otherwise, skip to Frame 16.

Adding Numbers

13. Addition is the means by which two or more numbers are combined and expressed as a single quantity called a sum or total. The plus (+) sign is used to denote addition.

When a group of numbers is to be added, units are added to units, tens are added to tens, hundreds are added to hundreds, and so on. Therefore they should be aligned so that whole numbers and decimals are properly placed in relation to one another.

Examples: (a) 3 + 45 = 48 or

```
   3
+ 45
----
  48  sum or total
```

(b) To add 3.5, .006, 1265, 9.0965, line up the columns exactly as they are shown here and add each column.

```
   3.5
    .006
1265.
   9.0965
---------
1277.6025  sum or total
```

When the total of any column is greater than one digit, the additional digits are added or "carried" to the next column to the left. For example, in the following problem,

```
 2.16
13.094
 5.176
28.039
 1  21
------
48.469
```

4 + 6 + 9 = 19. Record as 9 in the thousandths column and carry the "1" to the hundredths column. Then, 1 + 3 + 7 + 9 + 6 = 26. Record 6 in the answer and carry the "2" to the tenths position. The total of the tenths column is 4. Proceed in the same manner for the remaining columns. You may "carry" numbers mentally or write them down as illustrated.

14. Write the following numbers in a column and add: 375, 80, .05, 2.65, 82.009.

- - - - - - - - - - - - - - -

```
375.
 80.
   .05
  2.65
 82.009
    1
------
539.709
```

15. Write in a column and add the following: seven-tenths, nine hundred sixty, five thousandths, and three and five-tenths.

- - - - - - - - - - - - - - -

```
   .7
960.
   .005
  3.5
  1
-------
964.205
```

Developing Speed and Accuracy in Addition
Grouping by Tens

16. Do you use combinations often when you add? If so, try these problems. Otherwise, go to Frame 17.

Add these columns, using combinations of 10 when possible.

(a)
```
524
619
790
297
713
624
___
```

(b)
```
872.15
250.08
852.22
 96.55
180.09
 21.01
______
```

(a)
```
 524
 619
 790
 297
 713
 624
 22
____
3567
```

(b)
```
  872.15
  250.08
  852.22
   96.55
  180.09
   21.01
  3 1 1 3
 _______
 2272.10
```

Did you use all these combinations of ten? If so, go on to Frame 19. If you would like more practice, continue here.

17. Look for and combine two or more numbers that add to 10 in the same column. In the example shown below, note that 7 and 3 add to 10 and that 6 plus 4 equals 10 in the units column. Add the remaining numbers in that column. The result is 24. The digit 2, which is carried to the ten's position, combines with 8 to equal 10, to which other digits in that column are added.

```
 37
 53
 46
 91
 84
 93
 2
___
404
```

(Writing down the number carried makes your addition easier to check if you are interrupted or need to re-add.)

Other number combinations that tend to repeat themselves, such as 12, 20, 25, can be used in the same way to increase your speed and accuracy in adding.

18. Find the sum of each of the following, using combinations of ten whenever possible. Show combinations of tens as illustrated in the preceding frame.

(a)
```
   8. 3 6
   1. 0 3
   9. 2 1
 1 6. 1 4
 ________
```

(b)
```
 1 9
   6
 2 5
 7 1
 ___
```

(c)
```
 1 6
   8
 2 0
 7 2
   9
 2 2
 ___
```

(d)
```
 1. 0 2
 3. 2 5
 4. 5 3
  . 1 8
 5. 2 2
 ______
```

(a)
```
   8. 3 6┐     ┐
  ┌1. 0 3┤     │
  └9. 2 1┘ or  │
 1 6. 1 4      ┘
 2      1
 ________
 3 4. 7 4
```

(b)
```
┌1 9┐
│  6│
├2 5┘
└7 1┘
  2
 _____
 1 2 1
```

(c)
```
┌1 6   ┐
│  8┐  │
├2 0│or│
└7 2┘  │
   9   │
 2 2   ┘
 2
 _____
 1 4 7
```

(d)
```
   1. 0 2┐
   3.┌2 5┤
  ┌4.├5 3┘
  │ .├1 8┐
  ├5.└2 2┘
  └1  2
 ________
 1 4. 2 0
```

Developing Speed and Accuracy in Addition Using Subtotals

19. When a long column of numbers is to be added, first add part of the column and record to the side, and then add these partial totals (subtotals) to get a final total. This method is helpful, since you may be interrupted, in which case the whole column need not be re-added.

Example:
```
  328
  422
  511   1261 subtotal
  ----  ----
  741
  108
   62
 1550   2461 subtotal
 ----   ----
  145
   22
 2100
  914
   62   3243 subtotal
 ----   ----
 6965   6965 total
```

20. In the example just shown, the subtotal 1261 is the total of the first three numbers, 2461 is the total of the next four numbers, and 3243 is the total of the last __________ numbers.

- - - - - - - - - - - - - - -

5

21. In the following problems, record the subtotals and totals as indicated.

(a)		(b)	
36		3.17	
127		22.09	
890		55.43	________
3110	________	9.44	
346		8.69	
578		83.75	________
289	________	2.09	
64		18.26	________ total
22	________ total		

- - - - - - - - - - - - - - -

(a)	(b)
4163	80.69
1213	101.88
86	20.35
5462	202.92

Checking Answers

22. The reverse-order method of checking answers in addition is the most common. This means that if you have added a column of numbers from the top down, you re-add by starting from the bottom and adding up. Add this column and check by the reverse-order method.

(b) Add from the bottom to the top and record total here. ____

2365
821
9067
544
8178
975
1819

Obviously, if your two totals are not the same, you have made an error.

(a) Add from the top down and record total here. ____

- - - - - - - - - - - - - - -

23,769

Crossfooting: Vertical and Horizontal Addition

23. One definition of the word "foot" is to add (a column of figures) and set down a total. To crossfoot means to add horizontally and set down a total.

Many business reports require both vertical and horizontal addition. Data are arranged to provide more than one kind of information, besides including a check against totals.

The following report provides (a) the total sales for each district for the week and (b) the total sales for each day of the week. With the data arranged this way, it is also easy to make comparisons between districts, days, or weekly periods during the year. The sum of the line or district totals ($1327.02) is also the sum of the column or daily totals and is referred to as the grand total.

MANLY CATERING SERVICE

Week Ending June , 19

Sales District	Monday	Tuesday	Wednesday	Thursday	Friday	Saturday	Total
A	$ 43.20	$ 92.52	$ 78.69	$ 86.98	$ 99.35	$120.50	$ 521.24
B	71.15	76.56	68.25	91.00	82.52	99.20	488.68
C	55.17	66.42	45.98	54.13	45.00	50.40	317.10
Total	$169.52	$235.50	$192.92	$232.11	$226.87	$270.10	$1327.02

24. In the example above, $169.52 is a (district/daily) total __________

$317.10 is a (district/daily) total __________

The grand total is $__________

The sum of the district totals is $__________

The sum of the daily totals is $__________

- - - - - - - - - - - - - -

daily

district

$1327.02

$1327.02

$1327.02

Since the sum of the district totals equals the sum of the daily totals, the report is assumed to be correct. (It is possible but rare to have a compensating error.)

Identifying and Correcting Errors in Crossfooting

25. You can save a great deal of time if you know how to locate an error. The possibilities depend on the type of problem. In general, proceed as follows:

A. Find the difference between the sum of the line totals and the sum of the column totals.
   (1) If the difference appears in the units column, re-add that column only; if it appears in the tens column, re-add the first two columns, and so on.
   (2) If the difference equals any one number in the report, it is likely that it was omitted in adding.
   (3) An error of 1, 10, or any power of 10 is almost always an error in addition rather than in copying or omitting a number.

B. If the error cannot be localized in these ways, re-add the total column and verify the line totals.

C. As a last resort check each column total and then check each line total.

Use this procedure to find the error in the following report. Draw a line through the incorrect total and record the correct amount directly above it. Record the correct grand total.

A & C BUILDERS

Contracts for July

Type of Contract	District			
	A	B	C	Total
1-A	36	77	49	162
3-Z	109	82	76	267
2-C	25	56	40	121
Total	170	205	165	___

(a) The sum of the total line as shown is ________.

(b) The sum of the total column as shown is ________.

(c) The correct grand total is ________.

- - - - - - - - - - - - - -

(a) 540, (b) 550, (c) 550; total of column B should be 215.

26. One common fault in copying numbers is transposing digits--or reversing their order. When this occurs, the difference from the control figures will be a multiple of 9 (e.g., 9, 18, 27).

    If 36 is copied as 63, the difference will be ______. Is this a multiple of 9? ________. If the digits in the number 91 are transposed when copied, the number will be ______. The difference will be ______ which is a multiple of ____.

    - - - - - - - - - - - - - - -

    27, yes, 19, 72, 9

27. These figures were copied from other records. One number has been transposed. Add each column and line. Then compare your totals with the correct totals shown at the right. Locate the error, draw a line through it, and record the correct number directly above it. Correct the totals affected by this error in the same manner.

A	B	C	Total	A	B	C	Total
36	98	32	______				157
57	27	49	______				133
19	38	51	______				108
				112	154	132	398

- - - - - - - - - - - - - - -

Answer should appear as follows:

A	B	C	Total
	89		157
36	~~98~~	32	~~166~~
57	27	49	133
19	38	51	108
	154		398
112	~~163~~	132	~~407~~

Notice that in each case the difference between the correct and incorrect figures was 9.

28.

A & B MERCHANTS				
Sales Report		Week Ending January	, 19	
	Department			Total
	A	B	C	
Anza	$ 2,316.94	$ 7,546.18	$ 5,627.93	$15,491.05
Ballard	8,140.06	10,092.22	7,901.01	26,133.29
Manor	9,200.10	8,654.55	7,598.66	25,453.31
Hayes	5,476.18	7,078.00	3,950.75	16,504.93
Total	$25,133.28	$33,370.85	$25,078.35	$83,582.58

The sum of the column totals equals $83,582.48, a difference of 10 cents from the sum shown in the total column. This suggests an error in addition. Since it is easier to add vertically than horizontally, begin by checking the department totals, adding just the first two columns of digits (cents columns). Find it? Good for you.

The error was in Department ____. The total should read

$________________.

\- - - - - - - - - - - - - - - -

B; $33,370.95

29. When errors are a multiple of 10, it suggests an error in

__________________ rather than an omission or a number copied incorrectly. When the error cannot be identified, it is easier to check the (horizontal/vertical) addition first.

\- - - - - - - - - - - - - - - -

addition; vertical

Estimating Answers

30. An approximation is a value that is nearly but not exactly correct. There are many occasions, such as in comparative studies, decision making, and checking against gross errors in computational work, when exact figures are neither desirable nor necessary.

The report of stock market sales for any day may be 12 million shares when actually 12,426,429 shares may have been sold. The population of a certain city is reported to be 350,000, whereas 345,263 may be the exact number. In both cases the approximate figures are more meaningful to the general public. They also mean more when comparisons are made.

Although the exact answer is required in a business transaction when a settlement is made, rough estimates are used by many people in such trades as carpentry, repair work, and plumbing when bidding on a job.

The ability to estimate the answer to a problem can prevent a serious error, since the approximation serves as a guide to the correct solution. With a little practice you can develop skill in detecting errors that otherwise might be missed.

The key to estimating answers in addition is to round the numbers. In the following frames we will review rounding of numbers. If you think you can bypass this review, try the problems below. Otherwise go to Frame 31.

(a) Round to hundreds: 206, 375, 1090

(b) Round to units: 160.25, 0.578, 16.499

(c) Round to thousands: 906,096, 13,499, 27,500

(d) Round to tenths: 0.621, 12.068, 10.049

(e) Round to thousandths: 0.0783, 10.1009, 4.02

- - - - - - - - - - - - - -

(a) 200; 400; 1100
(b) 160; 1; 16
(c) 906,000; 13,000; 28,000
(d) 0.6; 12.1; 10.0
(e) 0.078; 10.101; 4.020

If you made any errors in these problems or had any difficulty, continue to Frame 31. Otherwise skip ahead to Frame 41.

Rounding Numbers

31. "Rounding" is an approximation of a number. How many digits remain after rounding (these digits are called significant figures) will depend on how you want to use the numbers, as we will see in later chapters. For the present, let us look at the basic rule in rounding numbers.

If the number to be dropped begins with 5 or more, add 1 to the preceding digit; if the number to be dropped begins with any digit less than 5, drop it and leave the rest of the number unchanged.

Examples:	Rounded to Hundreds	Rounded to Thousands
	863 to 900	821,862 to 822,000
	450 to 500	37,500 to 38,000
	319 to 300	9,499 to 9,000
	Rounded to Units	Rounded to Hundredths
	32.50 to 33	0.7349 to 0.73
	41.25 to 41	0.10409 to 0.10
	70.92 to 71	3.5 to 3.50
	6.499 to 6	44.0653 to 44.07

32. When 925 is rounded to the nearest hundreds it is 900, because the next digit, 2, is less than 5. In this case the digits 2 and 5 are dropped. Then, according to this rule, 916 rounded to the nearest hundreds equals ________ because the next digit, 1, is (equal to/ less than/more than) 5.

- - - - - - - - - - - - - - - -

900, less than

33. When 856 is rounded to the nearest hundreds it equals 900 because the second digit is 5. Therefore raise the digit 8 to 9. According to this rule, 857 rounded to hundreds equals ________ because the second digit is (equal to/less than/more than) 5.

- - - - - - - - - - - - - - - -

900, equal to

34. If 6.899 is rounded to the nearest whole number it equals 7, because the digit 8 is more than 5. Then 17.699 rounded to the nearest whole number is _____.

Round 8.52 to the nearest whole number __________

1678 to the nearest hundreds __________

2189 to the nearest thousands __________

- - - - - - - - - - - - - - - -

18, 9, 1700, 2000

35. Round the following numbers to the nearest hundreds:

217, 865, 1009

200, 900, 1000

36. Round the following numbers to the nearest thousandths: .0076, 31.1541, 16.05449.

.008; 31.154; 16.054

37. Round the following numbers to the nearest units or whole numbers: 56.5, 102.4, 7.499.

57, 102, 7

38. The number 1277.65 rounded to the nearest whole number is (1277/1278).

1278

39. 8249 rounded to the nearest hundreds is (8200/8300).

8200

40. The number 62.499 corrected or rounded to the nearest tenth is (62.5/62.4).

62.5

Estimating Answers in Addition

41. Suppose we wanted to know the approximate total of the following list of numbers

367, 125, 820, and 578

Consider the first digit only. Since these numbers are all 3-digit numbers, we round them to the nearest 100 (i.e., 400, 100, 800, and 600, respectively). Then

400 + 100 + 800 + 600 = 1900 estimated total

and 367 + 125 + 820 + 578 = 1890 exact total

What is your estimate of the following list? What is the actual total?

420, 106, 834, and 576

Estimated total = ______ + ______ + ______ + ______ = ______.

Exact total = 420 + 106 + 834 + 576 = ______.

- - - - - - - - - - - - - - -

Estimated total = 400 + 100 + 800 + 600 = 1900
Exact total = 1936

42. In we wished to estimate the total of the following column of numbers, we would find that the values differ considerably. A good general rule is to group them and round to the most common multiple of 10 (i.e., 100, 1000). Here we are using approximate subtotals to help us estimate the total.

Example:

Exact	Estimate
26	
840	900
35	
156	200
96	
805	900
212	
56	
8	300
2234	2300

Problem: What is your estimate of the following?

22
368
109
8
19
625
35
82
192

______ Exact ______ Estimate

- - - - - - - - - - - - - - -

Exact	Estimate
22	
368	400
109	100
8	
19	
625	600
35	
82	
192	300
1460	1400

OR

Exact	Estimate
22	
368	400
109	
8	
19	100
625	
35	700
82	
192	300
1460	1500

43. The total weekly operating costs of a certain business were $360.25, $1068.25, $8260.50, and $82.60. What is your estimate of the total? Ignore cents column; treat as $360, $1068, and so on.

Exact	Estimate
$ 360.25	
1068.25	
8260.50	
82.60	
$	$

- - - - - - - - - - - - - - -

Exact	Estimate
$ 360.25	$ 400
1068.25	1100
8260.50	8300
82.60	100
$9771.60	$9900

Unit 3

SUBTRACTING NUMBERS

44. Before beginning this topic, try these problems to see if you need to review subtracting and checking your answers. Subtract the following. Check answers mentally by addition--for example, 26 - 12 = 14; check: 14 + 12 = 26.

(a) 92 - 16 = ____

(b) 180 - 99 = ____

(c) 816.21
 - 36.09

(d) 186.21
 - 99.99

(e) Four thousand seven hundred six and five-hundredths minus four hundred ten and one hundred twenty-three thousandths.

(f) Eight hundred nine and seven-thousandths less ninety and ninety-eight hundredths.

- - - - - - - - - - - - - - - -

(a) 76, (b) 81, (c) 780.12, (d) 86.22

(e)
```
  4706.05
-  410.123
  4295.927
```

(f)
```
  809.007
-  90.98
  718.027
```

If you had any difficulty or made any errors, go on to Frame 45. If not, skip ahead to Frame 52.

45. Subtraction is the process of finding the difference between numbers. Unless otherwise stated, the minus (-) sign is used to indicate subtraction. Numbers are aligned as in addition so that units are subtracted from units, tens are subtracted from tens, etc., with whole numbers and decimals placed in proper position with respect to each other.

To check answers, add mentally the difference (answer) to the subtrahend, which should, of course, equal the minuend.

```
  36.76  minuend
- 20.75  subtrahend
  16.01  difference or remainder
```

Or 36.76 - 20.75 = 16.01
Check 16.01 + 20.75 = 36.76

46. Some problems are not as simple as the one shown in the preceding frame. Look at the following examples.

(a)
```
  45
- 18
  27
```

In this case, "8" in the units column is <u>greater than</u> "5." Since we cannot subtract 8 from 5, we use a system called "borrowing." If we borrow 10 from the tens column, we have 15 - 8 = 7. Since we borrowed 10 from the tens column, we must reduce the 4 to 3. We now subtract 1 from 3 and write the difference, which is 2. Our final answer is 27.

(b)
```
  3.12
- 1.0683
```

Zeros may be added to any number following the decimal point without changing its value.

Then our problem becomes

```
  3.1200
- 1.0683
  2.0517
```

and we proceed as in problem (a) by "borrowing" when necessary to obtain the difference.

47. Circle answers that are incorrect in the following examples:

(a) 489 − 237 = 252   (b) 101.56 − 39.59 = 61.87   (c) 31,377 − 18,788 = 12,689

(d) 631.45 − 238.831 = 392.619   (e) 4007 − 3110 = 997   (f) 504.18 − 231.07 = 263.11

(b), (c), (e), (f)

48. Subtract. Check results mentally.

(a) 61.53 − 49.02   (b) 50,677 − 33,786   (c) .00583 − .00505

(d) 10,017 − 9,009   (e) 21,054 − 11,009   (f) 234.56 − 88.05

(a) 12.51; (b) 16,891; (c) .00078; (d) 1008; (e) 10,045: (f) 146.51

49. Subtract. Check results mentally.

Subtract one thousand nine-hundred three and seventeen-hundredths from two thousand fifty-six and two-hundred nine thousandths.

2,056.209
- 1,903.17
153.039

50. Data may be arranged in reports so that figures are added in one direction and added or subtracted in the other direction. The business report shown on the next page represents dollar sales for the Havenscourt Services for the current year compared with the same period for the preceding year. Complete this report.

HAVENSCOURT SERVICES

Comparative Sales Report

	This Year	Last Year	Difference	
First 6 months	$2,316,820.08	$3,516,008.20	$ ________	decrease
Last 6 months	4,008,920.00	3,927,111.19	$ ________	increase
Total	$ ________	$ ________	$ ________	

- - - - - - - - - - - - - - -

This year: $6,325,740.08; Last year: $7,443,119.39; Difference: $1,199,188.12 decrease, $81,808.81 increase, $1,117,379.31 decrease.

51. Numerical information can also be arranged this way.

Find the net weight of the following carriers. (Gross weight equals the weight of the carrier plus the contents; tare weight equals the weight of the carrier; net weight equals the weight of the contents--gross weight less tare weight.) Using the principle of crossfooting, verify the total net weight for the five carriers listed.

In the following example: (a) Subtract the "Tare" weight from the "Gross" weight for each of the amounts shown and enter in the spaces provided. (b) Add each column. (c) Prove addition: Does the total of the "Net" column equal the total "Gross" column less the total of "Tare" column? (If it does, the addition is assumed to be correct.

	Gross	Tare	Net
	13,620	462	________
	27,317	462	________
	18,504	462	________
	35,614	462	________
	34,250	462	________
Total			

- - - - - - - - - - - - - - -

Net		
13,158	Gross total	129,305
26,855	Tare total	2,310
18,042	Net total	126,995
35,152		
33,788		
126,995		

Estimating Answers in Subtraction

52. Mr. Jones wanted to remodel his kitchen. One company offered to do the job for $876, and another company wanted $1052--a difference of approximately $200 ($1100 - $900). The actual difference was $176.

Exact		Estimate
$1052	minuend	$1100
876	subtrahend	900
$ 176	difference	$ 200

In estimating differences, round each number to as many places as there are digits in the smaller number (subtrahend).

Examples: (a) Exact Estimate (b) Exact Estimate

(a) Exact	Estimate	(b) Exact	Estimate
9836	9800	436.50	440
- 146	- 100	- 25.00	- 30
9690	9700	411.50	410

Estimate the differences for the following problems. Show your work as illustrated above.

(a) Exact	Estimate	(b) Exact	Estimate
658	________	768.50	________
- 43	________	- 29.42	________

difference

- - - - - - - - - - - - - - -

(a) 615, 660 - 40 = 620 (b) 739.08, 770 - 30 = 740

53. To arrive at a closer estimate when both numbers are of the same magnitude you may want to round to the second digit from the left rather than the first; for example, in the problem below you would round to the nearest 100 rather than 1000.

Example:

Exact	Estimate
2683	2700
- 1421	- 1400
1262	1300

Since both numbers are 4-digit numbers, they are not rounded to thousands but instead, to hundreds.

Estimate the difference for the following problems.

(a)

Exact	Estimate
7342	______
- 6250	______
1092	

(b)

Exact	Estimate
22106	______
- 19842	______
2264	

(c)

Exact	Estimate
8206	______
- 4310	______
3896	

- - - - - - - - - - - - - - - -

(a) 7300 - 6300 = 1000, (b) 22,000 - 20,000 = 2000,
(c) 8200 - 4300 = 3900

54. What is your estimate for the following?

(a)

Exact	Estimate
3168	______
- 236	______
2932	

(b)

Exact	Estimate
896.50	______
- 76.20	______
820.30	

(c)

Exact	Estimate
2963	______
- 1983	______
980	

- - - - - - - - - - - - - - - -

(a) 3200 - 200 = 3000, (b) 900 - 80 = 820, (c) 3000 - 2000 = 1000

SELF-TEST

Before you go to the next chapter take this Self-Test. Compare your answers with those given at the end of the test.

1. The following numbers are read as:

   236 ______________________________

   .617 ______________________________

   102.56 ______________________________

   .1060 ______________________________

   2,360,015 ______________________________

2. Write the following as numbers:

   Two thousand sixty-five and sixteen-hundredths ________

   One hundred twenty and nine-thousandths ________

   Two million, eighty-two thousand and six-tenths ________

   One thousand, two hundred fifty-eight ________

3. Add the following:

   (a) 20.67 + .009 + 2036.94 + 2.06

   (b) two and eight-tenths + fifty-five and nine-hundredths + two hundred four and sixteen-hundredths

4. Add by grouping tens:

(a)
```
      3. 1 6
    1 5. 4 2
      8. 0 4
  2 1 0. 9 8
  __________
```

(b)
```
    2 1 9
  3 0 0 6
    4 2 7
    1 1 5
    8 0 4
  _______
```

5. Add by use of subtotals:

(a)
```
2065
8174
9301  ________
 999
8006
 273  ________
```

(b)
```
15.72
 3.16  ________
 6.95
18.25  ________
83.21
 6.31  ________
```

6. Round the following numbers to the

(a) nearest whole number

26.95 ________

128.02 ________

52.39 ________

(b) nearest tenth

3084.06 ________

210.27 ________

99.1499 ________

(c) nearest thousandth

.07834 ________

1.6999 ________

(d) nearest hundredth

364.499 ________

57.0949 ________

7. What is the exact and approximate total of the following list of numbers? 360, 492, 1086, 2045, 965, 45, 715, 99

Exact　　　　　　　　Approximate

8. The number of square yards of carpet needed for several rooms in an office building is 68, 75, 14, 26, and 93. What are the exact and estimated totals?

Exact　　　　　　　　Estimate

9. Complete the following report. (Gross Sales less Returns and Allowances equals Net Sales.)

JAMES L. LAMPSON

Weekly Net Sales Report

Department	Gross Sales	Returns & Allowances	Net Sales
A	$1389.65	$205.16	$______
B	2009.25	39.82	______
C	3769.85	756.00	______
Total	$	$	$

10. Sales for the A & B Market dropped from $38,921.50 to $26,840.52 during the past two days. What are the exact and approximate differences?

Exact | Approximate

11. What are the exact and approximate differences between (a) $1376.50 and $799.25? Between (b) $832.17 and $56.80?

(a) Exact | Approximate | (b) Exact | Approximate

12. Arc Company bid $3486 on a contract while Mann Bros. bid $4409 on the same job. What are the actual and approximate differences between the two bids? (Estimate to nearest (a) thousands, (b) hundreds.)

Exact | (a) approximate | (b) approximate

Answers to Self-Test

Compare your answers to the Self-Test with the answers given here. If you miss any questions, review the frames indicated in parentheses before continuing to the next chapter. If you think you need more practice, additional problems are offered in the Appendix.

1. two hundred, thirty-six
   six hundred, seventeen thousandths
   one hundred two, and fifty-six hundredths
   one thousand sixty ten-thousandths
   two million, three hundred sixty thousand, fifteen (Frame 6A)

2. 2065.16
   120.009
   2,082,000.6
   1,258 (Frame 6B)

3.
```
(a)   20.67          (b)   2.8
        .009              55.09
    2036.94              204.16
       2.06              ------
    --------             262.05
    2059.679                        (Frame 12)
```

4.
```
(a)      3. 1 6       (b)   2 1 9
       1 5. 4 2           3 0 0 6
         8. 0 4             4 2 7
     2 1 0. 9 8             1 1 5
         1 1 2              8 0 4
     ----------           1   3
     2 3 7. 6 0           -------
                          4 5 7 1    (Frames 16-18)
```

5.
```
(a)  2,065            (b)  15.72
     8,174                  3.16   18.88
     9,301  19,540          6.95
       999                 18.25   25.20
     8,006                 83.21
       273   9,278          6.31   89.52
    ------  ------        ------  ------
    28,818  28,818        133.60  133.60   (Frame 19)
```

6. (a) 27
   128
   52

   (b) 3084.1
   210.3
   99.1

   (c) .078
   1.700

   (d) 364.45
   57.09 (Frame 30)

7. Exact

Exact	Approximate		
360		400	
492	900	500	
1086	1100	1100	
2045	2000	2000	
965		1000	
45	1000	OR	
715		700	
99	800	100	
253			
5807	5800	5800	(Frame 42)

8. Exact

Exact	Estimate	
68	70	
75	80	
14	10	
26	30	
93	90	
276	280	(Frame 41)

9. Gross Sales: \$7168.75; Returns & Allowances: \$1000.98; Net Sales: \$6167.77.

Departments:		
	A	\$1184.49
	B	1969.43
	C	3013.85
		\$6167.77 (Frames 23 and 50)

10. Exact: \$38,921.50 - \$26,840.52 = \$12,080.98
Approximate: \$40,000 - \$30,000 = \$10,000
or \$39,000 - \$27,000 = \$12,000 (Frame 52)

11. (a) 1376.50 − 799.25 = 577.25; 1400 − 800 = 600
(b) \$832.17 − 56.80 = \$775.37; \$830 − 60 = \$770 (Frame 52)

12. \$4409 − 3486 = \$ 923
(a) \$4000 − 3000 = \$1000
(b) \$4400 − 3500 = \$ 900 (Frame 53)

CHAPTER TWO

Multiplication and Division (Whole Numbers and Decimals)

OBJECTIVES

In this chapter you will learn to use shortcuts in:

- multiplying by any power of 10 (10, 100, 1000, etc.);
- multiplying by any factor of 10, 100, 1000, etc., such as 5, 25, 50, $16\frac{2}{3}$, 125;
- multiplying by 9, 99, 101, 90, 110, etc.;
- estimating answers in multiplication;
- dividing by any power of 10 (10, 100, 1000, etc.);
- dividing by any factor of 10, 100, 1000, etc., such as 5, 25, 50, $16\frac{2}{3}$, 125;
- estimating answers in division.

Unit 1

MULTIPLICATION

1. First, to see if you need to review multiplying and checking whole numbers and decimal numbers, try the following problems.

(a) 415 x 56

$$\begin{array}{r} 415 \\ \underline{56} \end{array}$$

(b) 910 x 12

$$\begin{array}{r} 910 \\ \underline{12} \end{array}$$

(c) 8.25 x .006

$$\begin{array}{r} 8.25 \\ \underline{.006} \end{array}$$

(d) 90 x 500

$$\begin{array}{r} 500 \\ \underline{90} \end{array}$$

(e) .062 x .005

$$\begin{array}{r} .062 \\ \underline{.005} \end{array}$$

(f) 83 x 1.25

$$\begin{array}{r} 1.25 \\ \underline{83} \end{array}$$

- - - - - - - - - - - - - - -

(a)		(b)		(c)	
	415		910		8.25
	56		12		.006
	2490		1820		.04950
	2075		910		
	23240		10920		

(d)		(e)		(f)	
	500		.062		1. 2 5
	90		.005		8 3
	45000		.000310		3 7 5
					1 0 0 0
					1 0 3.7 5

If you made mistakes or had any difficulty, continue to Frame 2.
If not, skip to Frame 8.

Multiplying Numbers

2. Multiplication is a short method of repeat addition; for example, 3 multiplied by 4 equals 12, which is the same as 3 added 4 times (3 + 3 + 3 + 3) or 4 added 3 times (4 + 4 + 4).

The result of multiplying two numbers (called factors) by one another is called the product. Since the result is the same, regardless of the factor used as the multiplier, it is easier to use the smaller number.

In the problem 3 x 6 = 18, 3 and 6 are called ________________.

The result, 18, is called the ________________.

- - - - - - - - - - - - - - -

factors, product

3. We will illustrate, with examples, the multiplication of whole numbers with two or more digits.

Examples

(a) 357 x 7 — 357 can be thought of as equal to 300 + 50 + 7, then 357 x 7 = (300 + 50 + 7) x 7,

or 357	multiplicand	
7	multiplier	
49	7 x 7	partial product
350	7 x 50	partial product
2100	7 x 300	partial product
2499	final product	

This is a long drawn out process so instead, we proceed as follows:

```
 357
   7
2499
```

(1) 7 x 7 = 49. Write "9" and mentally carry the "4" to the next position.

(2) 7 x 5 = 35. 35 + 4 = 39. Write "9" and carry the "3."

(3) 7 x 3 = 21. 21 + 3 = 24. Since there are no more digits, write down 24. Final answer is 2499.

(b) 365 x 54

```
  365   multiplicand
   54   multiplier
 1460     365 x 4  partial product
1825      365 x 5  partial product
19710   answer (product)
```

Problem: Complete the following multiplication.

```
296
 32
```

- - - - - - - - - - - - - - -

```
 296
  32
 592
888
9472
```

4. We will now illustrate multiplying numbers with zeros. Zeros are treated like all other digits.

Example:

```
 2050
   32
 4100  =  2050 x 2
6150   =  2050 x 3
65600
```

Rule to Remember: Any number multiplied by zero = zero. Then:

Problem: Complete the following multiplication.

```
90073
   52
```

```
  90073
     52
 ------
 180146
450365
-------
4683796
```

5. To multiply decimal numbers with two or more digits, the product must have as many decimals as there are in both factors.

Example: Multiply 3.605 by 1.27.

```
  3.6 0 5   multiplicand (3 decimals)
    1.2 7   multiplier (2 decimals)
---------
  2 5 2 3 5
    7 2 1 0
  3 6 0 5
-----------
  4.5 7 8 3 5   answer or product (5 decimals)
```

Problem: Complete the following multiplication.

```
2.0 6 4
  1 3.2
-------
```

```
    2.0 6 4
      1 3.2
  ---------
    4 1 2 8
  6 1 9 2
2 0 6 4
-----------
2 7.2 4 4 8   answer or product
```

6. Indicate the position of the decimal point in the answers for the following problems:

(a) 3.16 x 20 = 6 3 2 0 (b) 52 x .0078 = 4 0 5 6

(c) 21.96 x 5.34 = 1 1 7 2 6 6 4 (d) 1.85 x 6.257 = 1 1 5 7 5 4 5

(e) .18 x .07 = 1 2 6

- - - - - - - - - - - - - - -

(a) 63.20; (b) .4056; (c) 117.2664; (d) 11.57545; (e) .0126

Checking Answers

7. Answers may be checked in multiplication work by (a) reversing the factors or (b) dividing the product by either factor to obtain the other. Because multiplication is an easier process than division, the first method is usually preferred.

Example: If 3 x 40 = 120, then 40 x 3 = 120 or 120 ÷ 3 = 40.

Verify the answers to the following problems by reversing the factors.

```
(a)      3.2 5   Check:    5.6     (b)    1 3.7 5   Check:    2 3.9
           5.6            3.2 5             2 3.9            1 3.7 5
        ------           ------        ----------           -------
         1 9 5 0                          1 2 3 7 5
       1 6 2 5                            4 1 2 5
       -------                          2 7 5 0
       1 8.2 0 0                        ----------
                                        3 2 8.6 2 5
```

- - - - - - - - - - - - - - -

```
(a)        5.6                     (b)      2 3.9
          3.2 5                            1 3.7 5
       --------                         ----------
          2 8 0                            1 1 9 5
        1 1 2                            1 6 7 3
       1 6 8                             7 1 7
       --------                        2 3 9
       1 8.2 0 0                       ----------
                                       3 2 8.6 2 5
```

Multiplying by Any Power of 10 (10, 100, 1000, etc.)

8. Whole Numbers: If a number is to be multiplied by 10, 100, 1000, etc., simply add to the multiplicand the number of zeros in the multiplier.

   Examples: 96 x 10 = 960
   96 x 100 = 9600
   96 x 1000 = 96000

   Decimal Numbers: If the multiplicand is a decimal, move the decimal point to the right one place for each zero in the multiplier.

   Examples: 3.167 x 10 = 31.67
   3.167 x 100 = 316.7
   3.167 x 1000 = 3167

9. When a number is multiplied by 100, add _____ zeros to the multiplicand. 4608 x 100 = ______________; 960 x 100 = ______________.

   - - - - - - - - - - - - - - -

   2; 460,800; 96,000

10. 162 x 10 = ______________; 36 x 1000 = ______________; 126 x 100 = ______________.

    - - - - - - - - - - - - - - -

    1620; 36,000; 12,600

11. If 16.5 is multiplied by 10, move the decimal point _____ place(s) to the right to obtain the product.

    2.64 x 10 = 26.4; 8.25 x 10 = __________; .19 x 10 = __________.

    - - - - - - - - - - - - - - -

    one; 82.5; 1.9

12. 80.5 x 10000 = ______________; 10.645 x 100 = ______________; .0368 x 1000 = ______________.

    - - - - - - - - - - - - - - -

    805,000; 1064.5; 36.8

Multiplying by Any Factor of 10, 100, 1000, etc., such as 5, 25, 50, 16 2/3, and 125

13. (a) 629 x 5

    629 x 10 = 6290
    6290 ÷ 2 = 3145 (answer)

    Multiply by 10 and then divide the result by 2 since 10 is 2 times as great as 5.

    (b) 45 x 25

    45 x 100 = 4500
    4500 ÷ 4 = 1125 (answer)

    Multiply by 100 and then divide the result by 4 since 100 is 4 times as great as 25.

    (c) 562 x 50

    562 x 100 = 56,200
    56,200 ÷ 2 = 28,100 (answer)

    Multiply by 100 and then divide the result by 2 since 100 is 2 times as great as 50.

    (d) $36 \times 16\frac{2}{3}$

    36 x 100 = 3600
    3600 ÷ 6 = 600 (answer)

    Multiply by 100 and then divide the result by 6 since 100 is 6 times as great as $16\frac{2}{3}$.

    (e) 912 x 125

    912 x 1000 = 912,000
    912,000 ÷ 8 = 114,000 (answer)

    Multiply by 1000 and then divide the result by 8 since 1000 is 8 times as great as 125.

Study these examples and then test yourself with the following statements.

(a) An easy way to multiply by 25 is to multiply by ______ and then divide by ______.

(b) To multiply 725 by 20, add _____ zeros and then divide by _____.

(c) An easy way to multiply a number by 50 is to multiply by _____ and then divide by _____.

- - - - - - - - - - - - - - - -

(a) 100, 4; (b) 2, 5; (c) 100, 2

14. Perform the following calculations, using shortcut methods.

(a) 16 x 5 = 160 ÷ 2 = __________

27 x 50 = 2700 ÷ 2 = __________

16 x 25 = 1600 ÷ 4 = __________

48 x 125 = 48,000 ÷ 8 = __________

(b) 316 x 50 = __________ ÷ _____ = __________

82 x 125 = __________ ÷ _____ = __________

126 x 5 = __________ ÷ _____ = __________

3.28 x 25 = __________ ÷ _____ = __________

(c) 96.4 x 125 = __________ ÷ _____ = __________

8.55 x 5 = __________ ÷ _____ = __________

15.24 x 50 = __________ ÷ _____ = __________

218 x 25 = __________ ÷ _____ = __________

- - - - - - - - - - - - - - - -

(a) 80; 1350; 400; 6000
(b) 31,600 ÷ 2 = 15,800; 82,000 ÷ 8 = 10,250; 1260 ÷ 2 = 630; 328 ÷ 4 = 82
(c) 96,400 ÷ 8 = 12,050; 85.5 ÷ 2 = 42.75; 1524 ÷ 2 = 762; 21,800 ÷ 4 = 5450

Multiplying by 9, 99, 101, 90, 110

15. Numbers such as these (there are many others) can be multiplied quickly by some other number. Any number may be expressed as the sum or difference of two numbers; for example,

9 has the same value as 10 - 1
99 has the same value as 100 - 1
101 has the same value as 100 + 1
90 has the same value as 100 - 10
110 has the same value as 100 + 10

Therefore the following examples may be solved as illustrated.

26 x 9 = 26 x (10 - 1) or 26(10 - 1) = 260 - 26 = 234

546 x 99 = 546(100 - 1) = 54,600 - 546 = 54,054

214 x 101 = 214(100 + 1) = 21,400 + 214 = 21,614

763 x 90 = 763(100 - 10) = 76,300 - 7630 = 68,670

3.54 x 110 = 3.54(100 + 10) = 354 + 35.4 = 389.4

16. Perform the following calculations, using shortcut methods.

(a) 36 x 9 = 360 - 36 = ________

29 x 99 = 2900 - 29 = ________

75 x 90 = 7500 - 750 = ________

8.7 x 101 = 870 + 8.7 = ________

123 x 110 = 12,300 + 1230 = ________

(b) 3.42 x 90 = ________ - ________ = ________

132 x 99 = ________ - ________ = ________

96.2 x 101 = ________ + ________ = ________

42.5 x 110 = ________ + ________ = ________

387 x 9 = ________ - ________ = ________

- - - - - - - - - - - - - - -

(a) 324; 2871; 6750; 878.7; 13,530
(b) 342 - 34.2 = 307.8; 13,200 - 132 = 13,068;
9620 + 96.2 = 9716.2; 4250 + 425 = 4675; 3870 - 387 = 3483

Estimating Answers in Multiplication

17. As in addition and subtraction, it is easy, with a little practice, to estimate answers in multiplication. An estimate tells us if our answer is approximately correct, or in some cases that it is incorrect. The closer your estimate to the true answer, the better, in checking errors.

Example: 562 x 43 = 24,166

Round each factor to the same number of digits. In this example round the number 43 first (since it has only two digits) to the nearest tens and then round the larger factor to the same place. Then, 560 x 40 = 22,400.

Problems: What is your estimate and actual answer for the following?

(a) 413 x 26 = ________ Estimate ______ x ______ = ________

(b) 857 x 75 = ________ Estimate ______ x ______ = ________

(c) 853 x 47 = ________ Estimate ______ x ______ = ________

- - - - - - - - - - - - - - -

(a) 10,738; Estimate: 410 x 30 = 12,300
(b) 64,275; Estimate: 860 x 80 = 68,800
(c) 40,091; Estimate: 850 x 50 = 42,500

18. Sometimes an allowance or adjustment has to be made in rounding one factor or the other to obtain a good approximation.

Example: 836 x 403 = 336,908

If both are rounded to hundreds, then 800 x 400 = 320,000. A closer estimate can be made mentally, however, by rounding 836 to 840 and 403 to 400; then 840 x 400 = 336,000.

Problems: What is your estimate of the following?

(a) 875 x 403 = 352,625 Estimate ______ x ______ = ________

(b) 907 x 585 = 530,595 Estimate ______ x ______ = ________

- - - - - - - - - - - - - - -

(a) 880 x 400 = 352,000;
(b) 900 x 600 = 540,000 or 900 x 590 = 531,000

19. When multiplying decimals, such as 32.56 x 5.80, drop the decimal point and treat them as whole numbers.

Example: 32.56 x 5.80 = 188.848; treat as 33 x 6, then 30 x 6 = 180 (estimate).

Problem: What is your estimate of 93.06 x 25.86 = 2406.5316?

Estimate: ________ x ________ = ____________

What other shortcut method might you use to make this estimate?

- - - - - - - - - - - - - - -

90 x 30 = 2700; 9300 ÷ 4 = 2325

20. When any number is multiplied by a decimal, the product will always be smaller than the multiplicand. This fact will help you avoid a serious error in checking your answer.

Examples: 32 x .016 = .512
12.6 x .24 = 3.024*
8.65 x .507 = 4.38555*

Problems: (a) 812 x .234 = __________

(b) 2.016 x .15 = __________

(c) 75.2 x .018 = __________

(a)
```
      8 1 2
      .2 3 4
    -------
    3 2 4 8
  2 4 3 6
1 6 2 4
-----------
1 9 0.0 0 8
```

(b)
```
   2.0 1 6
       .1 5
 ---------
 1 0 0 8 0
 2 0 1 6
 ---------
 .3 0 2 4 0
```

(c)
```
    7 5.2
   .0 1 8
  -------
  6 0 1 6
   7 5 2
  -------
 1.3 5 3 6
```

Unit 2

DIVISION

21. To see if you need any practice or review on dividing whole numbers and decimals, and checking your answers, try the following problems.

Whole Numbers: Show remainder. Check your answers by multiplication.

(a) 21 ⟌5683 (b) 702 ⟌73654 (c) 132 ⟌8206

check: check: check:

---

*If you know the decimal equivalents of common fractions, you will know that the answer for the second example will be slightly less than one-fourth of 12.6, and that the answer for the third example will be slightly more than one-half of 8.65. We will discuss these later.

Decimal Numbers: Find the quotient to the following to the nearest tenth.

(d) Divide 7.5 by 2.054

(e) Divide 4368 by 9.7

Find the quotient in the following, and correct it to the nearest hundredth.

(f) $526 \div 22$

(g) $33.92 \div 1.6$

(a)
```
     270
 21 |5683
     42
     148
     147
      13
```
check: (270 x 21) + 13 = 5683

(b)
```
        104
 702 |73654
      702
       3454
       2808
        646
```
check: (104 x 702) + 646 = 73654

(c)
```
         62
 132 |8206
      792
       286
       264
        22
```
check: (132 x 62) + 22 = 8206

(d)
```
                3.6  = 3.7
 2.0 5 4 |7.5 0 0 0
          6 1 6 2
          1 3 3 8 0
          1 2 3 2 4
            1 0 5 6
```

(e)
```
          4 5 0.3  = 4 5 0.4
 9.7 |4 3 6 8.0 0
      3 8 8
        4 8 8
        4 8 5
            3 0 0
            2 9 1
                9
```

```
(f)         2 3.9 0 9  = 2 3.9 1      (g)        2 1.2  = 2 1.2 0
     2 2 |5 2 6.0 0 0                      1.6 |3 3.9 2
          4 4                                   3 2
            8 6                                   1 9
            6 6                                   1 6
            2 0 0                                   3 2
            1 9 8                                   3 2
                2 0 0
```

If you had any errors or if you want some practice in dividing, continue to Frame 22. Otherwise skip to Frame 31.

Dividing Whole Numbers

22. Division is a short method of repeat subtraction. It is the inverse operation of multiplication; for example, 20 divided by 5 is the same as saying: "How many times can the number 5 be subtracted from 20?" In this case the number is 4.

Any division problem may be expressed as follows:

$362 \div 15$, $\frac{362}{15}$, or $15\overline{)362}$

(a) The number 14 can be subtracted from 70 five times. Then 70 divided by 5 = _____ and 70 divided by 14 = _____.

(b) 24 ÷ 6 is the same as asking how many times 6 can be _______________________ from 24. The answer is _____.

- - - - - - - - - - - - - - - -

(a) 14, 5; (b) subtracted, 4

23. The ability to divide quickly and accurately is dependent on a thorough knowledge of the multiplication tables. For example, 24 ÷ 6 = 4 because 6 x 4 = 24. What is 33 ÷ 8? We want to know how many 8s there are in 33. There are four 8s in 33 with a remainder of 1.

Find the answer to the following. Show the remainder as illustrated.

Example: 47 ÷ 9 = 5, R 2

(a) 56 ÷ 8 = ____________ (b) 87 ÷ 9 = ____________

(c) 37 ÷ 4 = ____________ (d) 17 ÷ 5 = ____________

(e) 52 ÷ 7 = ____________ (f) 55 ÷ 6 = ____________

(a) 7; (b) 9, R 6; (c) 9, R 1; (d) 3, R 2; (e) 7, R 3; (f) 9, R 1

24. When the answer to a division problem cannot be determined by observation, the process is longer and is referred to as long division.

Example:

```
               24   quotient
divisor   15 |362   dividend
              30
               62
               60
                2   remainder
```

(a) Since there are two digits in the divisor, we look at the first two digits in the dividend which are 36 and ask ourselves how many 15s are there in 36. The answer is 2.

(b) 2 x 15 = 30. Write down 30; subtract from 36. The difference is 6.

(c) Write down the remainder 6 and the next digit in the dividend which is 2.

(d) We now want to know how many 15s are there in 62. There are 4 since 4 x 15 = 60. Write down 60 and subtract from 62. Bring down the remainder 2. The answer is 24 with a remainder of 2.

Problems:

(a) 57 |9090 (b) 26 |735

(a)
```
      159
 57 |9090
     57
     339
     285
      540
      513
       27
```

(b)
```
       28
 26 |735
     52
     215
     208
       7
```

25. Divide 5764 by 83.

```
      69
 83 |5764
     498
      784
      747
       37
```

- In this case the first two digits are less than 83 so we ask ourselves how many 83s are there in 576. To find the answer to this question, we ask ourselves how many 8s are there in 57. The answer is 7. But 7 x 83 = 581 which is more than 576. Therefore 6 is our answer.
- We repeat this process in determining how many 83s there are in 784. 78 ÷ 8 = 9 with a remainder of 6 so we "try" 9. 9 x 83 = 747. Bring down the remainder of 37.

Divide the following showing the remainder as illustrated above.

(a) 76 |20596   (b) 37 |12568

(a)

```
       271
 76 |20596
     152
      539
      532
        76
        76
```

(b)

```
       339
 37 |12568
     111
      146
      111
       358
       333
        25
```

26. Sometimes zeros cause us trouble in dividing. Zeros are treated like any other digit. Remember than any number multiplied by zero equals zero, and zero divided by any number equals zero.

Examples:

(a)

```
      2001
 18 |36025
     36
      025
       18
        7
```

Bring down the zero as illustrated. Since 0 ÷ 18 = 0, enter 0 in the quotient and bring down the next digit 2. Again, since 2 is less than 18, enter another 0 in the quotient. Bring down the next digit which is 5. 25 ÷ 18 = 1 with a remainder of 7.

(b)

```
      2048
25 ⟌51217
    50
     121
     100
      217
      200
       17
```

In this case we find that 12 is less than 25. Therefore, enter 0 in the quotient and bring down the next digit.

Problems:

(a) 12 ⟌204008 (b) 35 ⟌706309

- - - - - - - - - - - - - - - -

(a)

```
      17000
12 ⟌204008
    12
     84
     84
      008
```

(b)

```
      20180
35 ⟌706309
    70
     63
     35
     280
     280
        9
```

Dividing Decimal Numbers

27. After the decimal point has been determined proceed as with dividing whole numbers. When the number of decimals in the divisor is equal to or less than those in the dividend, mentally move the decimal point in the dividend as many places as there are decimal places in the divisor to locate the decimal point in the quotient.*

---

*The dividend and divisor or, in fraction form, the numerator and denominator, may be multiplied by any number without changing its value; that is,

$$\frac{.2815 \times 100}{.05 \times 100} = \frac{28.15}{5} = 5.63$$

Examples:

$$\begin{array}{r} 3. \\ 4.001\overline{)12.003} \end{array} \qquad \begin{array}{r} 20.6+ \\ 136\overline{)2805.1} \end{array} \qquad \begin{array}{r} 5.63 \\ .05\overline{).2815} \end{array}$$

$$\begin{array}{r} .09 \\ .009\overline{).00081} \end{array} \qquad \begin{array}{r} .05 \\ .018\overline{).00090} \end{array} \qquad \begin{array}{r} .3 \\ 40.01\overline{)12.003} \end{array}$$

<u>When the number of decimals in the divisor is greater than in the dividend</u>, add decimal places in the dividend to equal those in the divisor. Zeros added to any number after the decimal point do not change its value.

Examples:

(a) Divide 1026 by .15

$$\begin{array}{r} 6840. \\ .15\overline{)1026.00} \end{array}$$

(b) Divide 80.26 by 2.016

$$\begin{array}{r} 39.+ \\ 2.016\overline{)80.260} \end{array}$$

Indicate the position of the decimal point in the quotient for the following, as illustrated in the example. Add zeros to the dividend when necessary.

Example: $.26\overline{)36.05^{\cdot}6}$

(a) $36.5\overline{)29.0078}$

(b) $.15\overline{)7563}$

(c) $23\overline{)7.896}$

(d) $.52\overline{)9580.46}$

(e) $.036\overline{)21.6}$

(f) $4.27\overline{)369}$

(g) $2.017\overline{)8.25}$

(h) $26\overline{).00789}$

(i) $.0036\overline{)78.01}$

- - - - - - - - - - - - - - -

(a) $36.5\overline{)29.0^{\cdot}078}$

(b) $.15\overline{)7563.00^{\cdot}}$

(c) $23\overline{)7^{\cdot}.896}$

(d) $.52\overline{)9580.46^{\cdot}}$

(e) $.036\overline{)21.600^{\cdot}}$

(f) $4.27\overline{)369.00^{\cdot}}$

(g) $2.017\overline{)8.250^{\cdot}}$

(h) $26\overline{)^{\cdot}.00789}$

(i) $.0036\overline{)78.0100^{\cdot}}$

28. Divide as indicated. Show remainder.

(a) $19\overline{)4.205}$ (b) $3.55\overline{)21.65}$

(c) $.42\overline{).0076}$ (d) $.026\overline{)28.4}$

(e) $2.014\overline{)82.6}$

(a)
$$\begin{array}{r} .221 \\ 19\overline{)4.205} \\ \underline{38} \\ 40 \\ \underline{38} \\ 25 \\ \underline{19} \\ 6 \end{array}$$

(b)
$$\begin{array}{r} 6. \\ 3.55\overline{)21.65} \\ \underline{2130} \\ 35 \end{array}$$

(c)
$$\begin{array}{r} .01 \\ .42\overline{).0076} \\ \underline{42} \\ 34 \end{array}$$

(d)
$$\begin{array}{r} 1092. \\ .026\overline{)28.400} \\ \underline{26} \\ 240 \\ \underline{234} \\ 60 \\ \underline{52} \\ 8 \end{array}$$

(e)

```
             4 1.
2.0 1 4 | 8 2.6 0 0
          8 0 5 6
            2 0 4 0
            2 0 1 4
                2 6
```

Checking Answers in Division

29. Multiply the quotient by the divisor and add the remainder (if any). The result should equal the dividend.

Example: 953 ÷ 22 = 43 with a remainder of 7

Check: 43 x 22 = 946; 946 + 7 = 953

If 615 ÷ 36 = 17 with a remainder of 3, then 17 x _____ = _____, and

_____ + 3 = 615.

- - - - - - - - - - - - - - - -

17 x 36 = 612, and 612 + 3 = 615

30. Divide 3828 by 255 and verify. Show work in space provided.

- - - - - - - - - - - - - - - -

```
       15 with a remainder of 3
255 | 3828
```

check: (255 x 15) + 3 = 3828

Rounding Answers in Division to a Specified Number of Places

31. Many times we are asked to calculate the answer to the nearest tenth, hundredth, thousandth, etc., rather than using the remainder as shown in preceding frames.

Examples:

(a) Divide 926.83 by 24 and record answer correct to the nearest hundredth.

```
       3 8.6 1
2 4 | 9 2 6.8 3
      7 2
      2 0 6
      1 9 2
        1 4 8
        1 4 4
            4 3
            2 4
            1 9
```

The remainder, 19, is equal to more than $\frac{1}{2}$ of 24. This means that the next digit in the quotient (if division is carried to the next place) will equal 5 or more. Knowing this, we do not bother to continue but automatically record the answer as 38.62.

Rule: When the next digit is 5 or more, increase the preceding digit by 1; if less than 5, leave as is.

(b) Divide 8.069 by 52 to nearest thousandth.

```
       .1 5 5
5 2 | 8.0 6 9
      5 2
      2 8 6
      2 6 0
        2 6 9
        2 6 0
            9
```

9 is less than $\frac{1}{2}$ of 52, therefore drop it. Answer is .155.

(c) Divide 2368 by 15 to nearest whole number.

```
      157
15 | 2368
     15
      86
      75
     118
     105
      13
```

13 is more than $\frac{1}{2}$ of 15, therefore the answer is 158.

Problems:

(a) Divide 5310 by 25 to the nearest whole number.

25 | 5310

(b) Divide 5310 by 25 to the nearest tenth.

25 | 5310

(c) Divide 980.65 by 273 to nearest thousandth.

273 | 980.65

(a)

```
      212   Answer: 212
 25 | 5310
      50
       31
       25
        60
        50
        10
```

(b)

```
        2 1 2.4   Answer: 212.4
 2 5 | 5 3 1 0.0
       5 0
         3 1
         2 5
           6 0
           5 0
           1 0 0
           1 0 0
```

(c)

```
       3.5 9 2
273 | 9 8 0.6 5 0
      8 1 9
      -----
      1 6 1 6
      1 3 6 5
      -------
        2 5 1 5
        2 4 5 7
        -------
          5 8 0
          5 4 6
          -----
            3 4
```

Dividing by Any Power of 10 (10, 100, 1000, etc.)

32. Whole numbers: Mark off as many decimal places as there are zeros in the divisor.

Examples: 54 ÷ 10 = 5.4
216 ÷ 100 = 2.16
29 ÷ 1000 = .029

Decimal numbers: Move the decimal point to the left as many places as there are zeros in the divisor.

Examples: 2.19 ÷ 10 = .219
436.54 ÷ 100 = 4.3654
92.7 ÷ 1000 = .0927

Problems: Divide 3168 by 10 ______________
3168 by 100 ______________
3168 by 1000 ______________

- - - - - - - - - - - - - - -

316.8; 31.68; 3.168

33. Divide 417.55 by 10 ______________
46.789 by 100 ______________
358.090 by 1000 ______________

- - - - - - - - - - - - - - -

41.755; .46789; .358090

Dividing by Any Factor of 10, 100, 1000, etc., Such as 5, 25, 50, 16 2/3, and 125

34. (a) 962 ÷ 5

962 ÷ 10 = 96.2 96.2 x 2 = 192.4	Divide by 10 and then multiply the result by 2, since 10 is 2 times as great as 5.

(b) 3620 ÷ 25

3620 ÷ 100 = 36.20 36.20 x 4 = 144.80	Divide by 100 and then multiply by 8, since 100 is 4 times as great as 25.

(c) 854 ÷ 125

854 ÷ 1000 = .854 .854 x 8 = 6.832	Divide by 1000 and then multiply by 8, since 1000 is 8 times as great as 125.

Study these examples and then answer the following questions.

35. To divide any number by 5, first divide by ________ and then multiply by ________.

- - - - - - - - - - - - - - -

10; 2

36. When a number is divided by 25, the result equals four times the amount when divided by ________.

- - - - - - - - - - - - - - -

100

37. Using the method just illustrated, divide 3680 by 12.5.

3680 ÷ ________ = 36.80; ______________ x 8 = ______________

- - - - - - - - - - - - - - -

3680 ÷ 100 = 36.80; 36.80 x 8 = 294.40

38. Work the following problems, using shortcut methods. Show remainder for those problems with uneven answers.

(a) 735 ÷ 25 = (735 ÷ ________) x _____ = ____________

(b) 5106 ÷ 100 = ______________

(c) 375 ÷ 1000 = ______________

(d) $2125 \div 125 =$ ______________________

(e) $3278 \div 5 =$ ______________________

- - - - - - - - - - - - - - - -

(a) $735 \div 25 = (735 \div 100) \times 4 = 7.35 \times 4 = 29.40$
(b) $5106 \div 100 = 51.06$
(c) $375 \div 1000 = .375$
(d) $2125 \div 125 = (2125 \div 1000) \times 8 = 2.125 \times 8 = 17.000$
(e) $3278 \div 5 = (3278 \div 10) \times 2 = 327.8 \times 2 = 655.6$

Arithmetic Mean (Averages)

39. Averages are used a great deal in making comparisons. Examples are: (1) the average income for a certain group of employees, (2) the average temperature in a certain area throughout the year or as between years during the same period, (3) the average daily attendance of school children during the school year.

The average or mean equals the sum of values in a series or group divided by the number of values. As a formula,

$$M = \frac{\Sigma X}{n}$$

where ΣX = sum of values in a series
n = number of values in the series

Look at the following examples:

(a) The temperatures in a certain city during a 5-day period were 92, 98, 88, 96, and 102 degrees, respectively. What was the average temperature?

Solution:

Sum of temperature values = 92 + 98 + 88 + 96 + 102 = 476
Average or mean = $476 \div 5 = 95.2$ degrees
Using the formula, $M = \frac{\Sigma X}{n} = \frac{476}{5} = 95.2$ degrees

(b) What is the average height for a group of girls of these heights: 4 ft, 10 in.; 4 ft, 1 in.; 4ft, 5 in.; 5 ft; 5 ft, 2 in.; and 5 ft?

Solution:

$4\frac{10}{12}$ the sum (Σ X) = $28\frac{1}{2}$ ft

$4\frac{1}{12}$ n = 6

$4\frac{5}{12}$ M = $28\frac{1}{2} \div 6 = 4\frac{3}{4}$ ft or 4 ft 9 in.

5

$5\frac{2}{12}$

5

$\frac{18}{12} = 1\frac{6}{12}$

$28\frac{6}{12} = 28\frac{1}{2}$

(If you don't remember how to add fractions, don't worry. That will be reviewed in Chapter 3.)

40. The A and C Manufacturing Company pays six typists the following monthly salaries: \$420, \$500, \$475, \$480, \$520, and \$400. What is the average salary paid?

ΣX = _______ + _______ + _______ + _______ + _______ + _______ =

\$__________

M = \$__________ ÷ _____ = \$__________

- - - - - - - - - - - - - - - -

ΣX = \$420 + \$500 + \$475 + \$480 + \$520 + \$400 = \$2795
M = \$2795 ÷ 6 = \$465.83

41. Mortgages held by the Personal Finance Company for a certain district are \$10,000, \$7000, \$5000, \$4000, \$30,300, \$8000, and \$12,000. What is the average mortgage held in the district?

ΣX = n = _____

M = \$__________ ÷ _____ = \$_______

- - - - - - - - - - - - - -

$$\begin{array}{r} \$10,000 \\ 7,000 \\ 5,000 \\ 4,000 \\ 30,300 \\ 8,000 \\ \underline{12,000} \\ \Sigma X = \$76,300 \end{array}$$

n = 7

M = $76,300 ÷ 7 = $10,900

Estimating Answers in Division

42. The procedure in estimating answers in division is simple, but it is usually more difficult to get a good approximation to the exact answer than it is in multiplication.

Estimates are generally more accurate if both the dividend and divisor are rounded up (Example a) or both are rounded down (Example b). Round the dividend to one more digit than the divisor.

Example (a): $38\overline{)4978}$ = 131

Example (b): $230\overline{)3220}$ = 14

Estimate: $40\overline{)5000}$ = 125

Estimate: $200\overline{)3000}$ = 15

43. When estimating an answer in division, if the divisor is rounded to 10s the dividend is rounded to 100s; if the divisor is rounded to 100s, the dividend is rounded to ________.

- - - - - - - - - - - - - -

1000s

44. Estimate the answers for the following:

Example: $86\overline{)4558}$ = 53

Estimate: $90\overline{)4600}$ = 51

Problem (a): $226\overline{)5198}$ = 23

Problem (b): $319\overline{)49,764}$ = 156

Estimate (a):

Estimate (b):

- - - - - - - - - - - - - - - -

(a) $5000 \div 200 = 25$; (b) $50,000 \div 300 = 167$, or $49,700 \div 320 = 155$

45. When estimating an answer in decimals for a problem such as $77.4656 \div 3.56 = 21.76$, round off decimals. Then, $77 \div 4 = 19\frac{1}{4}$. The estimate based on whole numbers is $80 \div 4 = 20$. Estimate the answer for $1389.0 \div 92.6$:

(a) Round to whole numbers ________ $\div$ ________

(b) Estimate is ________ $\div$ ________ = ________

(c) The exact answer is $1389.0 \div 92.6 =$ ________ (nearest whole number)

- - - - - - - - - - - - - - - -

(a) $1389 \div 93$; (b) $1400 \div 90 = 16$; (c) 15

SELF-TEST

Before you go to the next chapter take this Self-Test. Compare your answers with those given at the end of the test.

1. Multiply, using shortcuts when possible.

(a) 321 x 10 = ________ (b) .0078 x 52 = ________

(c) 42.6 x .18 = ________ (d) .19 x .016 = ________

(e) 500 x 297 = ________ (f) 56 x 125 = ________

(g) .017 x 100 = ________ (h) 99 x 378 = ________

(i) 38.75 x .25 = ________ (j) 101 x 84 = ________

(k) 138 x $16\frac{2}{3}$ = ________ (l) 1.865 x 1000 = ________

(m) 801 x 250 = ________ (n) .0087 x 100 = ________

(o) 19.8 x 50 = ________ (p) 986 x 5 = ________

(q) 75 x 9 = ________ (r) 427 x 90 = ________

(s) 416 x 110 = ________ (t) 24.6 x 9 = ________

2. Estimate the answers to the following.

(a) 467 x 14 = ________ (b) 56.7 x 12.2 = ________

(c) 908 x 27 = ________ (d) 82.06 x 9.3 = ________

(e) 857 x 85 = ________ (f) 16 x .95 = ________

3. Divide the following. Show remainder and verify your answer for (e) and (f) by multiplication. Use shortcuts when possible.

(a) 840 ÷ 50 = ________ (b) 9.06 ÷ 10 = ________

(c) 1000 ÷ 125 = ________ (d) 2.27 ÷ 100 = ________

(e) $218\overline{)620}$ check: 218 x ______ + ______ = 620

(f) $.21\overline{)32.78}$ check: .21 x _____ + _____ = 32.78

4. Record answers correct to the nearest hundredth.

   (a) $18\overline{)56}$ (b) $.084\overline{)7.38}$

   (c) $21\overline{)208.7}$ (d) $304\overline{)16000}$

5. Find the average daily attendance (nearest whole number) for the local theatre if the number of persons in attendance each day for a week were 300, 350, 375, 338, 526, 520, and 499.

6. Find the actual answer and estimate to the nearest whole number.

   (a) 248 ÷ 83 = __________ actual answer

   ______ ÷ ______ = __________ estimate

   (b) 92.54 ÷ 3.89 = __________ actual answer

   ______ ÷ ______ = __________ estimate

   (c) 592 ÷ 56 = __________ actual answer

   ______ ÷ ______ = __________ estimate

   (d) 106 ÷ 45.8 = __________ actual answer

   ______ ÷ ______ = __________ estimate

Answers to Self-Test

Compare your answers to the Self-Test with the answers given here. If you miss any questions, review the frames indicated in parentheses before continuing to the next chapter. If you think you need more practice, additional problems are offered in the Appendix.

1. (a) 3210 (b) .4056 (c) 7.668
   (d) .00304 (e) 148,500 (f) 7,000
   (g) 1.7 (h) 37,422 (i) 9.6875
   (j) 8484 (k) 23 (l) 1865
   (m) 200,250 (n) 0.87 (o) 990
   (p) 4930 (q) 675 (r) 38,430
   (s) 45,760 (t) 221.4 (Frames 8, 13, and 15)

2. (a) 470 x 10 = 4700 (b) 60 x 10 = 600, or 60 x 12 = 720
   (c) 900 x 30 = 27,000 (d) 80 x 10 = 800, or 80 x 9 = 720
   (e) 860 x 90 = 77,400 (f) 16 x 1 = 16
   (Frames 17 and 19)

3. (a) 16.8 (b) .906 (c) 8 (d) .0227
   (e) 2, R 184; check: (2 x 218) + 184 = 620
   (f) 156, R .02; check: (.21 x 156) + .02 = 32.78
   (Frames 24, 30, 32, and 34)

4. (a) 3.11 (b) 87.86 (c) 9.94 (d) 52.63
   (Frame 31)

5.
```
   300          n = 7
   350
   375          M = 2908 ÷ 7 = 415
   338
   526
   520
   499
ΣX = 2908
```
(Frames 39-41)

6. (a) actual 248 ÷ 83 = 2.9 = 3
   estimate 200 ÷ 80 = $2\frac{1}{2}$ = 3
   (b) actual 92.54 ÷ 3.89 = 24
   estimate 90 ÷ 4 = 23
   (c) actual 592 ÷ 56 = 11
   estimate 600 ÷ 60 = 10
   (d) actual 106 ÷ 45.8 = 2
   estimate 100 ÷ 50 = 2 (Frame 42)

CHAPTER THREE
Fractions

OBJECTIVES

In this chapter we will review how to work with fractions. When you complete the chapter you will be able to

- change fractions (proper, improper, simple, complex) and mixed numbers from one form to another;
- reduce fractions to lower terms and raise them to higher terms;
- add and subtract fractions using shortcuts to find the lowest common denominator;
- multiply and divide fractions using canceling, reducing, and common multiples or denominators as shortcuts;
- change a common fraction or mixed number to the decimal form, and vice versa.

Because much of the material in the first two units of this chapter is review material you may be able to bypass it altogether or to study only certain sections. The pretest below will help you identify which areas, if any, you should study.

After you take the test, check your answers with those at the end of the test and follow the instructions given there.

If you do not take the test but wish to review the entire chapter, go to Frame 1. If you are confident of your ability to work with fractions and wish to review only the shortcuts in Units 1 and 2, study pages 67-68, 73-74, 80, 83-84.

PRE-TEST

1. Which of the following are (a) proper fractions, (b) improper fractions, (c) mixed numbers, (d) complex fractions?

$\frac{3}{5}$ $\frac{21}{8}$ $\frac{3\frac{1}{2}}{2}$ $\frac{11}{12}$ $\frac{12}{11}$ $\frac{3}{9}$ $\frac{2\frac{1}{2}}{4}$ $\frac{6}{42\frac{1}{2}}$ $3\frac{1}{8}$ $\frac{21}{7}$ $5\frac{3}{7}$ $\frac{4\frac{1}{2}}{3\frac{1}{2}}$ $\frac{9}{4}$

(a) ____________________ (b) ____________________

(c) ____________________ (d) ____________________

2. Reduce the following fractions to the lowest terms:

(a) $\frac{12}{22} =$ (b) $\frac{33}{177} =$

(c) $\frac{20}{45} =$ (d) $\frac{27}{108} =$

3. Change each of the following improper fractions to a whole or mixed number reduced to its lowest terms:

(a) $\frac{22}{4} =$ (b) $\frac{327}{3} =$

(c) $\frac{17}{5} =$ (d) $\frac{110}{8} =$

4. Change each of the following mixed numbers to an improper fraction:

(a) $5\frac{2}{5} =$ (b) $3\frac{4}{5} =$

(c) $6\frac{1}{3} =$ (d) $12\frac{3}{4}$

5. Change each of the following fractions to higher terms as indicated:

(a) $\frac{2}{5} = \frac{\quad}{20}$ (b) $\frac{3}{8} = \frac{\quad}{40}$

(c) $\frac{4}{6} = \frac{\quad}{12}$ (d) $\frac{2}{9} = \frac{\quad}{27}$

6. Add the following:

(a) $2\frac{1}{3}$ (b) $12\frac{1}{6}$

$3\frac{2}{3}$ $8\frac{3}{20}$

$12\frac{1}{3}$ $4\frac{1}{5}$

6

(c) $10\frac{2}{3}$ (d) $10\frac{1}{3}$

$42\frac{1}{8}$ $5\frac{2}{7}$

$3\frac{5}{6}$ $126\frac{3}{5}$

$28\frac{1}{12}$

$4\frac{1}{9}$

7. Subtract the following:

(a) $26\frac{6}{8}$ (b) $520\frac{5}{9}$ (c) 616

$5\frac{3}{8}$ $19\frac{7}{9}$ $20\frac{3}{5}$

8. Multiply the following fractions:

(a) $\frac{9}{10} \times \frac{7}{3} =$ (b) $\frac{22}{8} \times \frac{2}{11} =$

(c) $16 \times \frac{5}{9} =$ (d) $\frac{2}{3} \times 12\frac{1}{2}$

(e) $3\frac{1}{3} \times 9\frac{3}{11}$

9. Divide the following fractions and reduce to lowest terms:

(a) $\frac{2}{5} \div \frac{1}{3} =$ (b) $\frac{8}{9} \div \frac{4}{12}$

(c) $25 \div 3\frac{1}{3} =$ (d) $15\frac{1}{2} \div 6\frac{1}{8} =$

(e) $8\frac{1}{6} \div 5\frac{1}{2} =$

10. Simplify (division):

(a) $\frac{4\frac{1}{3}}{22} =$

(b) $\frac{10}{3\frac{1}{2}} =$

(c) $\frac{5\frac{1}{6}}{8\frac{2}{7}} =$

(d) $\frac{\frac{2}{9}}{\frac{7}{6}} =$

Answers to Pre-test

After you check your answers, follow the instructions and review the frames indicated.

1. (a) $\frac{3}{5}, \frac{11}{12}, \frac{3}{9}$ (b) $\frac{21}{8}, \frac{12}{11}, \frac{21}{7}, \frac{9}{4}$

   (c) $3\frac{1}{8}, 5\frac{3}{7}$ (d) $\frac{3\frac{1}{2}}{2}, \frac{2\frac{1}{2}}{4}, \frac{6}{42\frac{1}{2}}, \frac{4\frac{1}{4}}{3\frac{1}{2}}$ (Frames 1-9)

2. (a) $\frac{6}{11}$; (b) $\frac{11}{59}$; (c) $\frac{4}{9}$; (d) $\frac{1}{4}$ (Frames 10-13)

3. (a) $5\frac{1}{2}$; (b) 109; (c) $3\frac{2}{5}$; (d) $13\frac{3}{4}$ (Frame 14)

4. (a) $\frac{27}{5}$; (b) $\frac{19}{5}$; (c) $\frac{19}{3}$; (d) $\frac{51}{4}$ (Frame 16)

5. (a) $\frac{8}{20}$; (b) $\frac{15}{40}$; (c) $\frac{8}{12}$; (d) $\frac{6}{27}$ (Frames 15, 17)

6. (a) $24\frac{1}{3}$; (b) $24\frac{31}{60}$; (c) $88\frac{59}{72}$; (d) $142\frac{23}{105}$ (Frames 1-11)

7. (a) $21\frac{3}{8}$; (b) $500\frac{7}{9}$; (c) $595\frac{2}{5}$ (Frames 12-16)

8. (a) $2\frac{1}{10}$; (b) $\frac{1}{2}$; (c) $8\frac{8}{9}$; (d) $8\frac{1}{3}$; (e) $30\frac{10}{11}$ (Frames 17-22)

9. (a) $1\frac{1}{5}$; (b) $2\frac{2}{3}$; (c) $7\frac{1}{2}$; (d) $2\frac{26}{49}$; (e) $1\frac{16}{33}$ (Frame 23)

10. (a) $\frac{13}{66}$; (b) $2\frac{6}{7}$; (c) $\frac{217}{348}$; (d) $\frac{4}{21}$ (Frames 24-25)

Instructions:

1. If you made no more than one error in any group of problems and not more than 5 errors for the entire test, skip to Unit 3. If you wish to review the shortcuts in Units 1 and 2, study pages 67-68, 73-74, 80, 83-84.
2. If you did not meet this requirement or if you felt unsure of your answers, begin with the frame indicated for the first section that gave you difficulty. Skip other sections where you made no errors.

Unit 1

CHANGING FRACTIONS OR MIXED NUMBERS TO ANOTHER FORM, REDUCING FRACTIONS, AND RAISING TO HIGHER TERMS

1. A fraction is another way of writing division; for example $\frac{9}{11}$ is called a fraction but it is also another way of saying that 9 is to be divided by 11.

   It may be expressed as a number (called the numerator) written above another number (called the denominator), such as $\frac{1}{2}$ and $\frac{3}{4}$ or in decimal form as .50 and .75, respectively.

   In the fraction $\frac{7}{8}$, 7 is called the ________________ and 8 is called the ________________.

   - - - - - - - - - - - - - - - -

   numerator, denominator

2. .875 is the ________________ form of $\frac{7}{8}$.

   - - - - - - - - - - - - - - - -

   decimal

Proper and Improper Fractions

3. A proper fraction has a value less than 1 because, by definition, it has a numerator (also called the dividend) that is less than the denominator (also called the divisor). Some examples are:

   $\frac{3}{8}$, $\frac{5}{6}$, $\frac{11}{12}$

An improper fraction has a value equal to or greater than 1 because, by definition, it has a numerator that is equal to or greater than the denominator. Some examples are:

$\frac{5}{4}, \frac{7}{6}, \frac{4}{3}, \frac{8}{8}$

(a) A proper fraction is one in which the numerator is (less than/greater than) the denominator.
(b) An improper fraction is one in which the numerator is (greater than or equal to/less than or equal to) the denominator.

- - - - - - - - - - - - - - -

(a) less than; (b) greater than or equal to

4. Group the following fractions as proper and improper fractions:

$\frac{2}{3} \quad \frac{9}{5} \quad \frac{11}{6} \quad \frac{4}{6} \quad \frac{12}{12} \quad \frac{3}{9} \quad \frac{7}{5} \quad \frac{5}{3} \quad \frac{1}{8}$

Proper fractions ______________________

Improper fractions ______________________

- - - - - - - - - - - - - - -

Proper fractions $\frac{2}{3}, \frac{4}{6}, \frac{3}{9}, \frac{1}{8}$; Improper fractions $\frac{9}{5}, \frac{11}{6}, \frac{12}{12}, \frac{7}{5}, \frac{5}{3}$

Mixed Numbers

5. A mixed number is a number containing a whole number and a fraction. Some examples are:

$3\frac{1}{2}, 4\frac{1}{4}, 8\frac{1}{5}$

If a mixed number contains a whole number and a fraction, then $2\frac{5}{6}$ is a ______________ number.

- - - - - - - - - - - - - - -

mixed

6. In a mixed number $2\frac{5}{6}$, $\frac{5}{6}$ is a ______________ fraction.

- - - - - - - - - - - - - - -

proper

7. Five and one-fourth written as a mixed number is ______.

- - - - - - - - - - - - - - -

$5\frac{1}{4}$

Simple and Complex Fractions

8. A simple fraction is one in which both the numerator and denominator are whole numbers, such as

$$\frac{9}{13} \quad \frac{5}{8} \quad \frac{6}{9} \quad \frac{8}{3} \quad \frac{5}{4}$$

A complex fraction is one in which the numerator or denominator, or both, is a mixed number. Some examples are

$$\frac{5\frac{1}{5}}{6\frac{1}{2}} \quad \frac{4}{2\frac{1}{3}} \quad \frac{10\frac{1}{2}}{5} \quad \frac{27\frac{2}{3}}{\frac{8}{9}}$$

In the fraction $\frac{2}{9}$ the numerator and denominator are both whole numbers. Therefore, $\frac{2}{9}$ is a ______________ fraction. It is also a ______________ fraction. Is $\frac{7}{4}$ a simple or a complex fraction? ______________. $\frac{7}{4}$ is also called a(n) ____________ fraction.

- - - - - - - - - - - - - - -

simple, proper, simple, improper

9. (a) Which of the following fractions are simple and which are complex?

$$\frac{12\frac{1}{2}}{4\frac{1}{4}} \quad \frac{2}{3} \quad \frac{5\frac{1}{2}}{3} \quad \frac{42}{42\frac{1}{4}} \quad \frac{14}{8} \quad \frac{5}{9} \quad \frac{3\frac{1}{3}}{10\frac{1}{2}} \quad \frac{4}{13}$$

Simple: ______________________________________

Complex: ______________________________________

(b) Which of the simple fractions listed under (a) are proper and which are improper?

Proper: ____________________ Improper: ________________

(a) Simple: $\frac{2}{3}$, $\frac{14}{8}$, $\frac{5}{9}$, $\frac{4}{13}$; Complex: $\frac{12\frac{1}{2}}{4\frac{1}{4}}$, $\frac{5\frac{1}{2}}{3}$, $\frac{42}{42\frac{1}{4}}$, $\frac{3\frac{1}{3}}{10\frac{1}{2}}$

(b) Proper: $\frac{2}{3}$, $\frac{5}{9}$, $\frac{4}{13}$; Improper: $\frac{14}{8}$

Reducing Fractions

10. To reduce a fraction to its lowest terms, divide the numerator and denominator by the largest whole number common to both.

Example: Reduce $\frac{12}{18}$ to its lowest terms. Since 6 is the largest divisor of 12 and 18, divide each by 6; that is,

$$\frac{12 \div 6}{18 \div 6} = \frac{2}{3}$$

If the largest common divisor or factor is not apparent, you will obtain the same result by dividing both the numerator and denominator of the fraction by any common divisor and repeating the process until the fraction is reduced to its lowest terms. In the example given above, 3 is a common divisor of both 12 and 18, in which case

$$\frac{12}{18} = \frac{12 \div 3}{18 \div 3} = \frac{4}{6}$$

Since both 4 and 6 may be divided by 2, then $\frac{4}{6} = \frac{2}{3}$. In this case, an additional step in the process is required but the final result is, of course, always the same.

11. In the fraction $\frac{21}{28}$ the largest factor that both the numerator and denominator may be divided by is _____. Then $\frac{21 \div \quad}{28 \div \quad}$ = _____.

7; $\frac{21 \div 7}{28 \div 7} = \frac{3}{4}$

12. When reducing or changing fractions to lower terms, the following suggestions may be helpful if you do not recognize common divisors of the numerator and denominator.

(a) When both the numerator and denominator are even, they can always be divided by 2 and sometimes by a multiple of 2.

Examples: $\frac{10}{12} = \frac{10 \div 2}{12 \div 2} = \frac{5}{6}$ $\quad$ $\frac{20}{24} = \frac{20 \div 4}{24 \div 4} = \frac{5}{6}$

(b) When either the numerator or denominator is even and the other is odd or both are odd numbers, they cannot be divided by an even number. If there is a common divisor, it must be an odd number such as 3, 5, or 7.

Examples: $\frac{21}{49} = \frac{21 \div 7}{49 \div 7} = \frac{3}{7}$ $\frac{18}{63} = \frac{18 \div 9}{63 \div 9} = \frac{2}{7}$

(c) When the numerator and denominator both end in 0 and/or 5, a common factor or divisor is 5 or a multiple of 5.

Examples: $\frac{20}{45} = \frac{20 \div 5}{45 \div 5} = \frac{4}{9}$ $\frac{85}{135} = \frac{85 \div 5}{135 \div 5} = \frac{17}{27}$

13. Reduce each of the following fractions to its lowest terms.

(a) $\frac{12}{20} =$ (b) $\frac{25}{45} =$

(c) $\frac{18}{36} =$ (d) $\frac{27}{81} =$

(e) $\frac{120}{225} =$ (f) $\frac{432}{1608} =$

(g) $\frac{258}{654} =$

(a) $\frac{12}{20} = \frac{12 \div 4}{20 \div 4} = \frac{3}{5}$ (b) $\frac{25}{45} = \frac{25 \div 5}{45 \div 5} = \frac{5}{9}$

(c) $\frac{18}{36} = \frac{18 \div 18}{36 \div 18} = \frac{1}{2}$ (d) $\frac{27}{81} = \frac{27 \div 27}{81 \div 27} = \frac{1}{3}$

(e) $\frac{120}{225} = \frac{120 \div 5}{225 \div 5} = \frac{24}{45} = \frac{24 \div 3}{45 \div 3} = \frac{8}{15}$, or $\frac{120}{225} = \frac{120 \div 15}{225 \div 15} = \frac{8}{15}$

(f) $\frac{432}{1608} = \frac{432 \div 8}{1608 \div 8} = \frac{54}{201} = \frac{54 \div 3}{201 \div 3} = \frac{18}{67}$, or $\frac{432}{1608} = \frac{432 \div 24}{1608 \div 24} = \frac{18}{67}$

(g) $\frac{258}{654} = \frac{258 \div 2}{654 \div 2} = \frac{129}{327} = \frac{129 \div 3}{327 \div 3} = \frac{43}{109}$, or $\frac{258}{654} = \frac{258 \div 6}{654 \div 6} = \frac{43}{109}$

For problems (e) and (f) it is not practical to look for factors as large as 15 and 24.

Changing an Improper Fraction to a Whole or Mixed Number

14. Divide the numerator by the denominator and write the remainder, if any, as a fraction.

Examples: $\frac{12}{7} = 1\frac{5}{7}$ mixed number (12 ÷ 7 = 1 with a remainder of 5)

$\frac{15}{5} = 3$ whole number (15 ÷ 5 = 3, no remainder)

Change the following improper fractions to a whole or mixed number.

(a) $\frac{22}{5} =$ (b) $\frac{14}{3} =$

(c) $\frac{21}{7} =$ (d) $\frac{124}{3} =$

(a) $4\frac{2}{5}$; (b) $4\frac{2}{3}$; (c) 3, (d) $41\frac{1}{3}$

Changing a Fraction to Higher Terms

15. The numerator and denominator of any fraction may be multiplied by the same number without changing its value. For example, $\frac{1}{2}$ and $\frac{2}{4}$ have the same value. In order to add or subtract fractions, it is often necessary to change them to higher terms. Follow the procedure as illustrated.

Examples: (a) $\frac{1}{4} = \frac{?}{20}$ (b) $\frac{3}{8} = \frac{?}{24}$

Procedure: 20 ÷ 4 = 5 24 ÷ 8 = 3

Answers: $\frac{1 \times 5}{4 \times 5} = \frac{5}{20}$ $\frac{3 \times 3}{8 \times 3} = \frac{9}{24}$

Now, using this procedure, work the following problem.

$\frac{15}{17} = \frac{?}{34}$ (a) 34 divided by 17 = ____

(b) then $\frac{15 \times \_\_\_\_}{17 \times \_\_\_\_} = \frac{\_\_\_\_}{34}$

(a) 2; (b) $\frac{15 \times 2}{17 \times 2} = \frac{30}{34}$

16. Change each of the following fractions to the higher terms shown:

(a) $\frac{1}{5} = \frac{}{20}$ (b) $\frac{5}{9} = \frac{}{27}$

(c) $\frac{2}{3} = \frac{}{12}$ (d) $\frac{3}{4} = \frac{}{24}$

- - - - - - - - - - - - - - - -

(a) $\frac{4}{20}$; (b) $\frac{15}{27}$; (c) $\frac{8}{12}$; (d) $\frac{18}{24}$

Changing a Mixed Number to an Improper Fraction

17. Multiply the whole number by the denominator of the fraction, add the numerator of the fraction and place over the denominator.

Examples: $4\frac{3}{4} = \frac{19}{4}$ $5\frac{1}{6} = \frac{31}{6}$

Solution: $4 \times 4 = 16$ $5 \times 6 = 30$
$16 + 3 = 19$ $30 + 1 = 31$

Complete the following:

(a) $5\frac{2}{3} = \frac{}{3}$ (b) $12\frac{1}{2} = \frac{}{2}$ (c) $3\frac{2}{7} = \frac{}{7}$

(d) $8\frac{3}{4} = \frac{}{4}$ (e) $5\frac{2}{7} = \frac{}{7}$ (f) $6\frac{3}{8} = \frac{}{8}$

- - - - - - - - - - - - - - - -

(a) $\frac{17}{3}$; (b) $\frac{25}{2}$; (c) $\frac{23}{7}$; (d) $\frac{35}{4}$; (e) $\frac{37}{7}$; (f) $\frac{51}{8}$

Unit 2

ADDITION, SUBTRACTION, MULTIPLICATION, AND DIVISION OF FRACTIONS

Adding Fractions When the Denominators Are the Same

18. To add fractions when the denominators are the same, add the numerators of the fractions and divide by the denominator. (In the example which follows, the result is $6 \div 4$ or $1\frac{2}{4}$.) Reduce the fractional part to its lowest terms (in the example, $\frac{2}{4} = \frac{1}{2}$), and write it in the answer. Then carry the whole number (1 in this example) to the units column and add as usual.

<u>Example</u>: $4\frac{1}{4}$ $\qquad \frac{1}{4} + \frac{2}{4} + \frac{3}{4} = \frac{6}{4}$

$5\frac{2}{4}$ $\qquad \frac{6}{4} = 1\frac{2}{4} = 1\frac{1}{2}$

$12\frac{3}{4}$

$22\frac{2}{4}$ or $22\frac{1}{2}$ (answer)

19. <u>One</u> fourth plus <u>two</u> fourths plus <u>three</u> fourths = <u>six</u> fourths, or

$$\frac{1}{4} + \frac{2}{4} + \frac{3}{4} = \frac{6}{4}$$

Then <u>one</u> sixth plus <u>three</u> sixths plus <u>five</u> sixths is the same as

(a) $\frac{1}{6} + \frac{3}{6} + \frac{5}{6} =$ _____ or _____ reduced to lowest terms.

(b) $\frac{5}{8} + \frac{3}{8} =$ _____ expressed as a whole number.

(c) $\frac{3}{12} + \frac{1}{12} + \frac{5}{12} = \frac{\quad}{12}$, or $\frac{\quad}{4}$ when reduced to lowest terms.

(a) $\frac{9}{6}$ or $\frac{3}{2} = 1\frac{1}{2}$; (b) 1; (c) $\frac{9}{12}$ or $\frac{3}{4}$

20. Add the following:

(a) $4\frac{2}{5}$, $18\frac{3}{5}$, $9\frac{1}{5}$ (b) $5\frac{1}{9}$, $14\frac{2}{9}$, 6 (c) $2\frac{1}{4}$, $5\frac{3}{4}$, $10\frac{2}{4}$

(a) $32\frac{1}{5}$; (b) $25\frac{1}{3}$; (c) $18\frac{1}{2}$

Adding Fractions When the Denominators Are Not the Same

21. Fractions with unlike denominators cannot be added until the denominators are changed to the same value. (1) Find the smallest number that can be divided by all of the denominators. This number is called the least common denominator (LCD). In the example below, the smallest number that can be divided by 6, 8, and 4 is 24. (2) Change the fractions so that each has the same denominator (in this case, 24) and add as illustrated.

Example:

$5\frac{1}{6}$	$\frac{4}{24}$
$20\frac{3}{8}$	$\frac{9}{24}$
$15\frac{3}{4}$	$\frac{18}{24}$
$41\frac{7}{24}$	$\frac{31}{24} = 1\frac{7}{24}$

22. (a) The smallest number that can be divided by 3, 4, and 2 is 12; 12 is called the ______________________________.

(b) The LCD of 2, 6, and 3 is _____.

(c) In the series 4, 6, 12, and 8, is the LCD 24 or 36? _____

- - - - - - - - - - - - - - - -

(a) least common denominator; (b) 6; (c) 24

23. Add:

(a)

$4\frac{2}{3}$ $\frac{\ }{6}$

$16\frac{1}{6}$ $\frac{\ }{6}$

$5\frac{2}{6}$ $\frac{\ }{6}$

(b)

$5\frac{1}{8}$

$16\frac{3}{4}$

$22\frac{1}{12}$

$56\frac{3}{8}$

$\frac{5}{12}$

(c)

$22\frac{1}{5}$

$10\frac{2}{5}$

$5\frac{3}{10}$

$4\frac{1}{4}$

$\frac{6}{10}$

$16\frac{2}{4}$

9

- - - - - - - - - - - - - - - -

(a) $26\frac{1}{6}$; (b) $100\frac{3}{4}$; (c) $68\frac{1}{4}$

Finding the Least Common Denominator for a Series of Numbers

24. When one denominator can be divided exactly by the others, the LCD can be found by inspection.

Example: In the fractions $\frac{2}{3}$ and $\frac{5}{6}$, the LCD is 6 since it can be divided exactly by 3 and 6.

In $\frac{1}{3}$, $\frac{2}{9}$, and $\frac{1}{18}$, the LCD is 18 since it can be divided exactly by 3, 9, and 18.

What is the LCD for $\frac{3}{5}$, $\frac{1}{15}$, $\frac{2}{30}$? ________

- - - - - - - - - - - - - - - -

30 since it is divisible by 5, 15, and 30

25. When numbers have no common denominator (or factor) except 1, they are said to be prime to one another. The numbers 7, 9, and 13 are prime to one another since they have no common denominator except 1. The series 4, 5, and 9 is prime for the same reason. When denominators are prime to one another, the least common denominator is their product.

Example: Find the LCD for the following fractions:

$\frac{5}{6}$, $\frac{1}{7}$, and $\frac{4}{5}$

Solution: Since 6, 7, and 5 are prime to one another, the LCD = $6 \times 7 \times 5 = 210$.

What is the LCD for $\frac{2}{3}$, $\frac{7}{11}$, and $\frac{1}{5}$? ____

- - - - - - - - - - - - - - - -

LCD = $3 \times 11 \times 5 = 165$

26. (a) The LCD for $\frac{2}{3}$, $\frac{5}{7}$, and $\frac{2}{4}$ is $3 \times 7 \times 4 =$ ____ because 3, 7, and 4 are ____________ to one another.

(b) The LCD for $\frac{1}{8}$, $\frac{5}{9}$, and $\frac{4}{5}$ is ______________________.

(c) The LCD for $\frac{3}{16}$ and $\frac{5}{8}$ is 16 because 16 is divisible by 8 and 16 exactly. Then the LCD for $\frac{3}{5}$ and $\frac{1}{10}$ is _____ and the LCD for $\frac{8}{11}$ and $\frac{2}{33}$ is _____.

- - - - - - - - - - - - - - - -

(a) 84, prime; (b) 8 x 9 x 5 = 360; (c) 10, 33

27. When the least common denominator is not apparent, follow the procedure outlined below.

Example: $\frac{4}{5}, \frac{7}{12}, \frac{2}{9}, \frac{5}{8}$

Procedure: 1. Arrange the denominators in a row.

```
    5     12     9     8
```

2. Divide the smallest number that can be divided into two or more denominators and bring down to the next row with any denominators that were not divided.

```
2 | 5     12     9     8
    5      6     9     4
```

3. Continue this process until there are no divisors.

```
2 | 5     12     9     8
2 | 5      6     9     4
3 | 5      3     9     2
    5      1     3     2
```

4. Multiply all divisors and the numbers in the last row.

2 x 2 x 3 x 5 x 1 x 3 x 2 = 360 LCD

In the series $\frac{3}{4}, \frac{5}{7}, \frac{2}{9}$, and $\frac{1}{6}$, the LCD is found as follows:

1. Arrange in this order:

```
2 | 4     7     9     6
3 | 2     7     9     3
    2     7     3     1
```

2. What is the smallest number that can be divided into two or more of these numbers? The answer is 2. Divide 4 and 6 by 2. Bring down the other numbers.
3. 9 and 3 are divisible by 3.
4. Since there are no more divisors, multiply 2 x 3 x 2 x 7 x 3 x 1 = 252 LCD.

28. Now use this method to find the LCD for the following series:

(a) $\frac{2}{7}, \frac{5}{6}, \frac{4}{8}, \frac{3}{9}, \frac{2}{3}$ (b) $\frac{1}{15}, \frac{2}{3}, \frac{1}{5}, \frac{2}{6}$

(a)

2	7	6	8	9	3
3	7	3	4	9	3
	7	1	4	3	1

LCD = 2 x 3 x 7 x 4 x 3 = 504

(b)

3	15	3	5	6
5	5	1	5	2
	1	1	1	2

LCD = 3 x 5 x 2 = 30

29. Add the following numbers: (1) Find the LCD, (2) Change each fraction to the LCD value, (3) Add as usual.

$3\frac{1}{3}$

$15\frac{1}{5}$

$27\frac{1}{2}$

$402\frac{3}{4}$

$98\frac{1}{10}$

Show work here:

3	5	2	4	10

LCD = ____________

Total = $546\frac{53}{60}$

2	3	5	2	4	10
5	3	5	1	2	5
	3	1	1	2	1

LCD = 2 x 5 x 3 x 2 = 60

Subtracting Fractions When the Denominators Are the Same

30. To subtract fractions when the denominators are the same, follow the procedure illustrated below.

Example: $12\frac{4}{9} - 5\frac{1}{9} = 7\frac{3}{9}$ or $7\frac{1}{3}$

Procedure: 1. Subtract $\frac{1}{9}$ from $\frac{4}{9}$ and reduce.

2. Subtract 5 from 12.

Subtract the following:

(a) $19\frac{7}{8} - 11\frac{2}{8}$

(b) $26\frac{4}{8} - 10\frac{2}{8}$

(a) $8\frac{5}{8}$; (b) $16\frac{2}{8}$ or $16\frac{1}{4}$

31. Now try these. Be sure to reduce fractions when possible:

(a) $\frac{7}{8} - \frac{2}{8} =$ ________

(b) $\frac{5}{9} - \frac{3}{9} =$ ________

(c) $\frac{8}{12} - \frac{5}{12} =$ ________

(d) $1\frac{4}{5} - \frac{2}{5} =$ ________

(a) $\frac{5}{8}$; (b) $\frac{2}{9}$; (c) $\frac{3}{12}$ or $\frac{1}{4}$; (d) $1\frac{2}{5}$

32. In the following examples, we will illustrate subtracting mixed numbers.

Example 1: $4\frac{2}{8} - 1\frac{7}{8}$

Procedure: In this example, $\frac{7}{8}$ cannot be subtracted from $\frac{2}{8}$ so we borrow 1 from the 4 and add it to $\frac{2}{8}$. $1 + \frac{2}{8} = 1\frac{2}{8} = \frac{10}{8}$. (See Frame 17 to review changing a mixed number to an improper fraction.) Then $4\frac{2}{8} = 3\frac{10}{8}$ and our problem can now be expressed as $3\frac{10}{8} - 1\frac{7}{8} = 2\frac{3}{8}$, or

$$\begin{array}{r} \overset{3}{\cancel{4}}\,\frac{\overset{10}{\cancel{2}}}{8} \\ -\ 1\frac{7}{8} \\ \hline 2\frac{3}{8} \end{array}$$

Reduce the 4 to 3 and change $1\frac{2}{8}$ to $\frac{10}{8}$ as shown here and subtract.

Example 2:

$$\begin{array}{r} 10\frac{17}{12} \\ \cancel{11}\frac{\cancel{5}}{\cancel{12}} \\ -\ 3\frac{7}{12} \\ \hline 7\frac{10}{12} = 7\frac{5}{6} \end{array}$$

Procedure: Since 7 is greater than 5, borrow 1 whole number from the minuend, change to 12ths and add $\frac{12}{12}$ to $\frac{5}{12}$. Then subtract $\frac{7}{12}$ from $\frac{17}{12}$.

Subtract the following:

(a)
$$\begin{array}{r} 23\frac{5}{12} \\ -\ 18\frac{9}{12} \\ \hline \end{array}$$

(b)
$$\begin{array}{r} 56\frac{1}{9} \\ -\ 40\frac{4}{9} \\ \hline \end{array}$$

- - - - - - - - - - - - - - - -

(a)
$$\begin{array}{r} 22\frac{17}{12} \\ \cancel{23}\frac{\cancel{5}}{\cancel{12}} \\ -\ 18\frac{9}{12} \\ \hline 4\frac{8}{12} = 4\frac{2}{3} \end{array}$$

(b)
$$\begin{array}{r} 55\frac{10}{9} \\ \cancel{56}\frac{\cancel{1}}{\cancel{9}} \\ -\ 40\frac{4}{9} \\ \hline 15\frac{6}{9} = 15\frac{2}{3} \end{array}$$

Subtracting Fractions When the Denominators Are Not the Same

33. To subtract fractions with unlike denominators, change to common denominators and subtract as usual.

Example:

$$\begin{array}{r|c} 90\frac{5}{6} & \frac{15}{18} \\ -\ 27\frac{1}{9} & \frac{2}{18} \\ \hline 63\frac{13}{18} & \frac{13}{18} \end{array}$$

Procedure: The LCD of 6 and 9 is 18. Therefore, change $\frac{5}{6}$ and $\frac{1}{9}$ to 18ths and proceed as illustrated.

Subtract: (a) $32\frac{7}{9}$ $-\ 5\frac{2}{3}$ (b) $27\frac{5}{8}$ $-\ 6\frac{3}{5}$

(a)

$32\frac{7}{9}$	$\frac{7}{9}$
$-\ 5\frac{2}{3}$	$\frac{6}{9}$
$27\frac{1}{9}$	$\frac{1}{9}$

(b)

$27\frac{5}{8}$	$\frac{25}{40}$
$-\ 6\frac{3}{5}$	$\frac{24}{40}$
$21\frac{1}{40}$	$\frac{1}{40}$

Multiplying Fractions

34. To multiply fractions, multiply the numerators by each other and write the result as the numerator of the answer. Then multiply the denominators by each other to get the denominator of the answer. Reduce the answer to its lowest terms.

Example: $\frac{3}{8} \times \frac{5}{7} = ?$

Procedure:

1. Multiply the numerators by each other (i.e., $3 \times 5 = 15$).
2. Multiply the denominators by each other (i.e., $8 \times 7 = 56$). We now have

$$\frac{3 \times 5}{8 \times 7} = \frac{15}{56}$$

3. Reduce $\frac{15}{56}$, if possible, to its lowest terms. In this case, 15 and 56 have no common factor or divisor.

What is $\frac{4}{9} \times \frac{5}{11}$? ______________________

$$\frac{4}{9} \times \frac{5}{11} = \frac{4 \times 5}{9 \times 11} = \frac{20}{99}$$

35. When a whole number is multiplied by a fraction, think of the whole number as an improper fraction and multiply as usual.

Example: $5 \times \frac{3}{8}$

Solution: Think of 5 as $\frac{5}{1}$. Then $5 \times \frac{3}{8} = \frac{5}{1} \times \frac{3}{8} = \frac{15}{8} = 1\frac{7}{8}$. When the result is an improper fraction, always change to a mixed number.

Multiply the following:

(a) $6 \times \frac{5}{11} =$

(b) $15 \times \frac{2}{7} =$

(a) $6 \times \frac{5}{11} = \frac{6 \times 5}{11} = \frac{30}{11} = 2\frac{8}{11}$; (b) $15 \times \frac{2}{7} = \frac{15 \times 2}{7} = \frac{30}{7} = 4\frac{2}{7}$

36. When a whole number is multiplied by a mixed number, change both factors to improper fractions and multiply as usual.

Example: $7 \times 3\frac{1}{8}$

Solution: Think of 7 as $\frac{7}{1}$. Change $3\frac{1}{8}$ to an improper fraction ($\frac{25}{8}$).

Then, $7 \times 3\frac{1}{8} = \frac{7}{1} \times \frac{25}{8} = \frac{175}{8} = 21\frac{7}{8}$.

The second step would normally be written as $7 \times \frac{25}{8}$ since the whole number is understood to have a denominator of 1.

When the result is an improper fraction, always change it to a mixed number.

Now try these problems:

(a) $12 \times 3\frac{5}{11} =$

(b) $8 \times 10\frac{5}{9} =$

(a) $12 \times 3\frac{5}{11} = 12 \times \frac{38}{11} = \frac{456}{11} = 41\frac{5}{11}$

(b) $8 \times 10\frac{5}{9} = 8 \times \frac{95}{9} = \frac{760}{9} = 84\frac{4}{9}$

37. Sometimes it is convenient to reduce a fraction before it is multiplied by another.

Example: $\frac{4}{10} \times \frac{7}{9} = ?$

Solution: In this case, $\frac{4}{10}$ can be reduced to $\frac{2}{5}$ since 4 and 10 have a common divisor of 2. Our problem now becomes

$$\frac{2}{5} \times \frac{7}{9} = \frac{14}{45}.$$

Try these problems:

(a) $\frac{6}{18} \times \frac{7}{15} =$

(b) $\frac{3}{5} \times \frac{4}{16} =$

(a) $\frac{6}{18} \times \frac{7}{15} = \frac{1}{3} \times \frac{7}{15} = \frac{7}{45}$; (b) $\frac{3}{5} \times \frac{4}{16} = \frac{3}{5} \times \frac{1}{4} = \frac{3}{20}$

38. Often the solution can be simplified by cancelling the common factors. The numerator and denominator are divided by a common factor.

Look at problem (a) in Frame 37 (i.e., $\frac{6}{18} \times \frac{7}{15}$). Here, $\frac{6}{18}$ was reduced by dividing both 6 and 18 by the common factor 6:

$$\frac{\overset{1}{\cancel{6}}}{\underset{3}{\cancel{18}}} \times \frac{7}{15} = \frac{1 \times 7}{3 \times 15} = \frac{7}{45}$$

This problem could also have been solved as follows:

$$\frac{\overset{\overset{1}{\cancel{2}}}{\cancel{6}}}{\underset{9}{\cancel{18}}} \times \frac{7}{\underset{5}{\cancel{15}}} = \frac{7}{45}$$

Here 6 and 15 were divided by the common factor 3. Then 2 and 18 were divided by the common factor 2. Usually number 1 as shown here is understood and is therefore not recorded.

Example: $\frac{\overset{5}{\cancel{10}}}{\underset{2}{\cancel{12}}} \times \frac{\cancel{6}}{\cancel{2}} = \frac{5}{2} = 2\frac{1}{2}$

12 and 6 are divided by the common factor 6. 10 and 2 are divided by the common factor 2.

Try these:

(a) $\frac{10}{15} \times \frac{3}{7} =$

(b) $\frac{7}{8} \times \frac{4}{11} =$

(a) $\frac{\overset{2}{\cancel{10}}}{\underset{\cancel{5}}{\cancel{15}}} \times \frac{\cancel{3}}{7} = \frac{2}{7}$

(b) $\frac{7}{\underset{2}{\cancel{8}}} \times \frac{\cancel{4}}{11} = \frac{7}{22}$

39. Solve the following:

(a) $\frac{10}{15} \times \frac{3}{11} =$

(b) $\frac{2}{16} \times \frac{3}{4} =$

(c) $\frac{11}{12} \times \frac{3}{5} \times \frac{6}{22} =$

(d) $6 \times \frac{8}{9} =$

(e) $\frac{3}{9} \times 2\frac{1}{3}$

(a) $\frac{\overset{2}{\cancel{10}}}{\underset{\cancel{3}}{\cancel{15}}} \times \frac{\cancel{3}}{11} = \frac{2}{11}$ or $\frac{\overset{2}{\cancel{10}}}{\underset{\cancel{5}}{\cancel{15}}} \times \frac{\cancel{3}}{11} = \frac{2}{11}$

(b) $\frac{\cancel{2}}{\underset{8}{\cancel{16}}} \times \frac{3}{4} = \frac{3}{32}$ or $\frac{\cancel{2}}{16} \times \frac{3}{\underset{2}{\cancel{4}}} = \frac{3}{32}$

(c) $\frac{\cancel{11}}{\underset{\underset{2}{\cancel{4}}}{\cancel{12}}} \times \frac{\cancel{3}}{5} \times \frac{\overset{3}{\cancel{6}}}{\underset{2}{\cancel{22}}} = \frac{3}{20}$ or $\frac{\cancel{11}}{\underset{2}{\cancel{12}}} \times \frac{3}{5} \times \frac{\cancel{6}}{\underset{2}{\cancel{22}}} = \frac{3}{20}$

(d) $\overset{2}{\cancel{6}} \times \frac{8}{\underset{3}{\cancel{9}}} = \frac{16}{3} = 5\frac{1}{3}$

(e) $\frac{\cancel{3}}{\underset{3}{\cancel{9}}} \times 2\frac{1}{3} = \frac{1}{3} \times \frac{7}{3} = \frac{7}{9}$ or $\frac{\cancel{3}}{9} \times \frac{7}{\cancel{3}} = \frac{7}{9}$

Dividing Fractions

40. To divide fractions, invert the divisor and proceed as in multiplication.

Examples: $\frac{1}{5} \div \frac{1}{3} = \frac{1}{5} \times \frac{3}{1} = \frac{3}{5}$

$$\frac{9}{10} \div \frac{1}{3} = \frac{9}{10} \times \frac{3}{1} = \frac{27}{10} = 2\frac{7}{10}$$

$$\frac{11}{12} \div \frac{7}{4} = \frac{11}{\underset{3}{\cancel{12}}} \times \frac{\cancel{4}}{7} = \frac{11}{21}$$

$$16 \div \frac{86}{9} = \overset{8}{\cancel{16}} \times \frac{9}{\underset{43}{\cancel{86}}} = \frac{72}{43} = 1\frac{29}{43}$$

$$21\frac{1}{2} \div 4 = \frac{43}{2} \times \frac{1}{4} = \frac{43}{8} = 5\frac{3}{8}$$

Try these:

(a) $\frac{2}{9} \div \frac{5}{18} =$

(b) $\frac{2}{5} \div \frac{7}{9} =$

(c) $21 \div \frac{7}{8} =$

(d) $\frac{3}{8} \div \frac{15}{2} =$

(e) $15\frac{1}{2} \div \frac{1}{2} =$

(a) $\frac{2}{9} \div \frac{5}{18} = \frac{2}{\cancel{9}} \times \frac{\overset{2}{\cancel{18}}}{5} = \frac{4}{5}$

(b) $\frac{2}{5} \div \frac{7}{9} = \frac{2}{5} \times \frac{9}{7} = \frac{18}{35}$

(c) $21 \div \frac{7}{8} = \overset{3}{\cancel{21}} \times \frac{8}{\cancel{7}} = 24$

(d) $\frac{3}{8} \div \frac{15}{2} = \frac{\cancel{3}}{\underset{4}{\cancel{8}}} \times \frac{\cancel{2}}{\underset{5}{\cancel{15}}} = \frac{1}{20}$

(e) $15\frac{1}{2} \div \frac{1}{2} = \frac{31}{\not{2}} \times \frac{\not{2}}{1} = 31$

41. Any complex fraction may be written and solved as illustrated in the preceding frame; that is,

$\frac{5\frac{1}{8}}{2\frac{1}{2}}$ is the same as $5\frac{1}{8} \div 2\frac{1}{2}$

Then $5\frac{1}{8} \div 2\frac{1}{2} = \frac{41}{8} \div \frac{5}{2} = \frac{41}{\not{8}_4} \times \frac{\not{2}}{5} = \frac{41}{20} = 2\frac{1}{20}$

Examples:

$\frac{5\frac{1}{2}}{6} = 5\frac{1}{2} \div 6 = \frac{11}{2} \times \frac{1}{6} = \frac{11}{12}$

$\frac{2\frac{1}{3}}{8\frac{1}{5}} = 2\frac{1}{3} \div 8\frac{1}{5} = \frac{7}{3} \times \frac{5}{41} = \frac{35}{123}$

Complete:

(a) $\frac{2\frac{1}{2}}{15\frac{1}{2}} = 2\frac{1}{2} \div 15\frac{1}{2} =$

(b) $\frac{6\frac{1}{3}}{12} =$

(a) $2\frac{1}{2} \div 15\frac{1}{2} = \frac{5}{\not{2}} \times \frac{\not{2}}{31} = \frac{5}{31}$

(b) $\frac{6\frac{1}{3}}{12} = 6\frac{1}{3} \div 12 = \frac{19}{3} \times \frac{1}{12} = \frac{19}{36}$

42. In some cases a common multiple or denominator may be used to advantage, but there is no saving of time unless the common multiple or divisor is readily recognized.

Examples of Use of Common Multiple:

(a) $\frac{4\frac{1}{2}}{10} = \frac{4\frac{1}{2} \times 2}{10 \times 2} = \frac{9}{20}$

The common multiple must be a number that will simplify the fraction by changing either the numerator or denominator or both (if necessary) to a whole number. In this example, the common multiple is 2 since $4\frac{1}{2} \times 2 = 9$.

If the numerator had been $4\frac{2}{3}$, the common multiple would be 3 since $4\frac{2}{3} \times 3 = 14$.

If the numerator had been $5\frac{1}{6}$, the common multiple would be 6 since $5\frac{1}{6} \times 6 = 31$, etc.

(b) $\frac{5\frac{1}{4}}{108} = \frac{5\frac{1}{4} \times 4}{108 \times 4} = \frac{21}{432} = \frac{7}{144}$

(c) $\frac{27}{3\frac{1}{3}} = \frac{27 \times 3}{3\frac{1}{3} \times 3} = \frac{81}{10} = 8\frac{1}{10}$

Example of Use of Common Denominator:

$$\frac{14}{3\frac{1}{2}} = \frac{28}{2} \div \frac{7}{2} = \frac{28}{\not{2}} \times \frac{\not{2}}{7} = 4$$

Think of 14 as 28 ÷ 2. Since the denominators are the same, the 2s will cancel out when the divisor is inverted and the problem becomes 28 ÷ 7 = 4, in which case the third step in the solution can be eliminated.

If the divisor had been $3\frac{1}{3}$ instead of $3\frac{1}{2}$, then think of 14 as 42 ÷ 3 in which case the problem would become

$$\frac{42}{3} \div \frac{7}{3} = 6$$

Complete the following:

(a) $\frac{5\frac{1}{3}}{16} =$

(b) $\frac{8}{7\frac{1}{4}} =$

- - - - - - - - - - - - - - -

(a) $\frac{5\frac{1}{3}}{16} = \frac{5\frac{1}{3} \times 3}{16 \times 3} = \frac{16}{48} = \frac{1}{3}$

(b) $\frac{8}{7\frac{1}{4}} = \frac{32}{4} \div 7\frac{1}{4} = \frac{32}{4} \div \frac{29}{4} = \frac{32}{29} = 1\frac{3}{29}$

or using a common multiple instead of a common denominator,

$\frac{8}{7\frac{1}{4}} = \frac{8 \times 4}{7\frac{1}{4} \times 4} = \frac{32}{29} = 1\frac{3}{29}$

Unit 3

DECIMALS OR DECIMAL FRACTIONS

43. If you can solve the following problems correctly, you may be able to skip part of this unit.

A. Change the following fractions or mixed numbers to the decimal form. Record to 2 decimals and carry fractions.

Example: $\frac{2}{3} = .66\frac{2}{3}$

(1) $\frac{13}{39} =$ (2) $10\frac{1}{12} =$

(3) $2\frac{7}{9} =$ (4) $\frac{15}{28} =$

(5) $\frac{5}{12} =$ (6) $36\frac{12}{13} =$

- - - - - - - - - - - - - - -

(1) $\frac{13}{39} = \frac{1}{3} = .33\frac{1}{3}$; (2) $10\frac{1}{12} = 10.08\frac{1}{3}$; (3) $2\frac{7}{9} = 2.77\frac{7}{9}$;

(4) $\frac{15}{28} = .53\frac{16}{28} = .53\frac{4}{7}$; (5) $\frac{5}{12} = .41\frac{2}{3}$; (6) $36\frac{12}{13} = 36.92\frac{4}{13}$

B. Change the following decimal numbers to the fractional form.

Example: $6.16\frac{2}{3} = 6\frac{1}{6}$

(1) $.205 =$ (2) $.55\frac{5}{9} =$

(3) $3.06\frac{1}{4} =$ (4) $21.55 =$

(5) $10.875 =$ (6) $.0375 =$

- - - - - - - - - - - - - - -

(1) $.205 = \frac{205}{1000} = \frac{41}{200}$; (2) $.55\frac{5}{9} = \frac{55\frac{5}{9}}{100} = \frac{5}{9}$; (3) $3.06\frac{1}{4} = 3 + \frac{6\frac{1}{4}}{100} =$

$3 + \frac{1}{16} = 3\frac{1}{16}$; (4) $21.55 = 21\frac{55}{100} = 21\frac{11}{20}$; (5) $10.875 = 10\frac{875}{1000} = 10\frac{7}{8}$

(6) $.0375 = \frac{375}{10,000} = \frac{3}{80}$

If you made no errors or had no difficulty with these problems, skip to Frame 54; otherwise continue on here.

44. In Units 1 and 2, we worked with fractions expressed in the more common form--that is, one number divided by another, such as $\frac{3}{8}$ or $\frac{12}{5}$. In this unit we will deal with fractions expressed as decimals. In Frame 7 of Chapter 1 we learned that

.3 = three tenths
.305 = three hundred five thousandths, etc.

In addition, three tenths may be written as $\frac{3}{10}$

three hundred five thousandths may be written $\frac{305}{1000}$

This means that decimals are another way of expressing fractions, and they are referred to as decimals or decimal fractions. (These terms may be interchanged.)

Examples: .75 = seventy-five hundredths = $\frac{75}{100} = \frac{3}{4}$

.16 = sixteen hundredths = $\frac{16}{100} = \frac{4}{25}$

2.5 = two and five tenths = $2\frac{5}{10} = 2\frac{1}{2}$ (mixed number or $\frac{5}{2}$ as an improper fraction)

Changing a Decimal to a Common Fraction or a Mixed Number

45. To change a decimal to a common fraction or mixed number, replace the decimal with the appropriate multiple of 10 (10, 100, 1000, etc.) and reduce to the lowest terms.

Examples: $.25 = \frac{25}{100} = \frac{1}{4}$ (.25 = 25 hundredths)

$.375 = \frac{375}{1000} = \frac{3}{8}$ (.375 = 375 thousandths)

Now complete these:

(a) $.125 = \frac{}{1000} = \frac{}{8}$ (b) $.02 = \frac{}{100} = \frac{}{50}$

- - - - - - - - - - - - - - -

(a) $.125 = \frac{125}{1000} = \frac{1}{8}$ (b) $.02 = \frac{2}{100} = \frac{1}{50}$

46. When the decimal is $.16\frac{2}{3}$, $.08\frac{1}{3}$, $.11\frac{1}{9}$, etc., proceed as follows:

$$.16\frac{2}{3} = \frac{16\frac{2}{3}}{100} = 16\frac{2}{3} \div 100 = \frac{\cancel{50}}{3} \times \frac{1}{\underset{2}{\cancel{100}}} = \frac{1}{6}$$

Then $.08\frac{1}{3} = \frac{8\frac{1}{3}}{100} = 8\frac{1}{3} \div 100 = \frac{\cancel{25}}{3} \times \frac{1}{\underset{4}{\cancel{100}}} = \frac{1}{12}$,

and $.11\frac{1}{9} = \frac{11\frac{1}{9}}{100} = 11\frac{1}{9} \div 100 = \frac{\cancel{100}}{9} \times \frac{1}{\cancel{100}} = \frac{1}{9}$.

Work these problems:

(a) $.33\frac{1}{3} = \frac{33\frac{1}{3}}{100} = \frac{100}{3} \times \frac{1}{100} = \frac{100}{300} = \frac{}{3}$

(b) $.66\frac{2}{3} = \underline{\hspace{2cm}} = \underline{\hspace{1.5cm}} \times \underline{\hspace{1.5cm}} = \underline{\hspace{1.5cm}} = \underline{\hspace{1.5cm}}$

(a) $\frac{1}{3}$; (b) $\frac{66\frac{2}{3}}{100} = \frac{\overset{2}{\cancel{200}}}{3} \times \frac{1}{\cancel{100}} = \frac{2}{3}$

47. Complete the following:

$$3.83\frac{1}{3} = 3\frac{83\frac{1}{3}}{100} = 3 + \frac{83\frac{1}{3}}{100} = \underline{\hspace{5cm}}$$

$\frac{83\frac{1}{3}}{100} = \frac{\overset{10}{\cancel{250}}}{3} \times \frac{1}{\underset{4}{\cancel{100}}} = \frac{10}{12} = \frac{5}{6}$; then $3 + \frac{5}{6} = 3\frac{5}{6}$

48. Convert 12.625 to fraction form:

$$12.625 = \underline{\hspace{6cm}}$$

$.625 = \frac{625}{1000} = \frac{5}{8}$; $12.625 = 12 + \frac{5}{8} = 12\frac{5}{8}$

49. Is $.83\frac{1}{3}$ equivalent (has the same value) to $\frac{5}{6}$, $\frac{5}{12}$, $\frac{7}{8}$? (Change $.83\frac{1}{3}$ to a fraction and compare.)

- - - - - - - - - - - - - - - -

$.83\frac{1}{3} = \frac{83\frac{1}{3}}{100} = \frac{250}{3} \times \frac{1}{100} = \frac{5}{6}$

50. Does $.125 = \frac{1}{8}$? yes___ no ___ Verify your answer ____________

Does $.675 = \frac{7}{8}$? yes___ no ___ Verify your answer ____________

- - - - - - - - - - - - - - - -

yes $(.125 = \frac{125}{1000} = \frac{1}{8})$; no $(.675 = \frac{675}{1000} = \frac{27}{40})$

51. Change .136 to a fraction and reduce to lowest terms.

- - - - - - - - - - - - - - - -

$.136 = \frac{136}{1000} = \frac{17}{125}$

Changing a Common Fraction or Mixed Number to a Decimal

52. If a decimal can be expressed as a common fraction or mixed number, then the reverse is true. For example,

$$\text{if } .75 = \frac{3}{4}, \text{ then } \frac{3}{4} = 3 \div 4 \text{ or } 4\overline{)3.00}^{\;.75}$$

To change a common fraction to a decimal, divide the numerator by the denominator to as many decimal places as desired. You may use a fraction at the end of a decimal instead of rounding off as in Examples (3) and (4).

To change a common fraction to a mixed number, change the fraction part to a decimal and record as in Examples (3), (4), and (5).

Examples: (1) $\frac{3}{8} = .375$

(2) $\frac{1}{16} = .0625$

(3) $5\frac{1}{6} = 5.166\frac{2}{3}$, or 5.167 rounded to third decimal

(4) $8\frac{1}{12} = 8.08\frac{1}{3}$, or 8.0833 rounded to fourth decimal

(5) $10\frac{3}{5} = 10.60$ or 10.6

53. Find the decimal form of the following (carry answers to three decimals).

(a) $\frac{3}{4} =$ (b) $\frac{13}{4} =$

(c) $4\frac{1}{8} =$ (d) $12\frac{1}{8} =$

(e) $\frac{26}{6} =$ (f) $2\frac{2}{5} =$

(a) .750; (b) $3\frac{1}{4} = 3.250$; (c) 4.125; (d) 12.125; (e) $4\frac{2}{6} = 4\frac{1}{3} =$ 4.333; (f) 2.400

54. If you know what <u>aliquot</u> parts are and how to use them, try the following problems to see if you can skip this section. Otherwise, continue with Frame 55.

(a) $160 \times .37\frac{1}{2} =$ (b) $40 \times .62\frac{1}{2} =$

(c) $45 \times .22\frac{2}{9} =$ (d) $200 \times .125 =$

(e) $48 \times .25 =$ (f) $12.84 \times .08\frac{1}{3} =$

(g) $54 \times .16\frac{2}{3} =$ (h) $3.66 \times .83\frac{1}{3} =$

(i) $96 \times .20 =$ (j) $4.60 \times .33\frac{1}{3} =$

(a) $\overset{20}{\cancel{160}} \times \frac{3}{\cancel{8}} = 60$; (b) $\overset{5}{\cancel{40}} \times \frac{5}{\cancel{8}} = 25$; (c) $\overset{5}{\cancel{45}} \times \frac{2}{\cancel{9}} = 10$; (d) $\overset{25}{\cancel{200}} \times \frac{1}{\cancel{8}} = 25$;

(e) $\overset{12}{\cancel{48}} \times \frac{1}{\cancel{4}} = 12$; (f) $\overset{1.07}{\cancel{12.84}} \times \frac{1}{\cancel{12}} = 1.07$; (g) $\overset{9}{\cancel{54}} \times \frac{1}{\cancel{6}} = 9$;

(h) $\overset{.61}{\cancel{3.66}} \times \frac{5}{\cancel{6}} = 3.05$; (i) $\overset{19.2}{\cancel{96}} \times \frac{1}{\cancel{5}} = 19.20$ or $19\frac{1}{5}$; (j) $4.60 \times \frac{1}{3} = 1.53\frac{1}{3}$

If you made no errors or had no difficulty, go on to the Self-Test at the end of this chapter. If you wish to review aliquot parts, continue with Frame 55.

55. In everyday business and personal use, we often need to multiply by decimals or decimal fractions. Many of these are equivalents of common fractions, so you can save yourself a lot of time if you learn to recognize them.

For example, 27 x $.33\frac{1}{3}$ would be a complicated multiplication problem unless you knew that $.33\frac{1}{3}$ was equal to $\frac{1}{3}$. Then the problem is simply 27 x $\frac{1}{3}$ = 9. Such decimal equivalents are also referred to as aliquot parts of 1 or $1.00. An aliquot part is any number that can be divided evenly into another. For example,

$$1 \div .33\frac{1}{3} = 3, \quad 1 \div .125 = 8, \text{ etc.}$$

The decimal equivalents of common fractions are used so frequently that they are worth memorizing. The following table lists the decimal equivalents of the most common fractions.

Decimal Equivalents of Common Fractions

3rds		5ths		8ths		12ths	
$\frac{1}{3}$	$.33\frac{1}{3}$	$\frac{1}{5}$	.20	$\frac{1}{8}$	.125	$\frac{1}{12}$	$.08\frac{1}{3}$
$\frac{2}{3}$	$.66\frac{2}{3}$	$\frac{2}{5}$	.40	$\frac{2}{8}$	.25	$\frac{2}{12}$	$.16\frac{2}{3}$
		$\frac{3}{5}$	.60	$\frac{3}{8}$	.375	$\frac{3}{12}$	.25
4ths		$\frac{4}{5}$	.80	$\frac{4}{8}$	.50	$\frac{4}{12}$	$.33\frac{1}{3}$
$\frac{1}{4}$	.25			$\frac{5}{8}$	.625	$\frac{5}{12}$	$.41\frac{2}{3}$
$\frac{2}{4}$	.50	6ths		$\frac{6}{8}$	.75	$\frac{6}{12}$	.50
$\frac{3}{4}$	.75	$\frac{1}{6}$	$.16\frac{2}{3}$	$\frac{7}{8}$	.875	$\frac{7}{12}$	$.58\frac{1}{3}$
		$\frac{2}{6}$	$.33\frac{1}{3}$			$\frac{8}{12}$	$.66\frac{2}{3}$
		$\frac{3}{6}$	.50			$\frac{9}{12}$	.75
		$\frac{4}{6}$	$.66\frac{2}{3}$			$\frac{10}{12}$	$.83\frac{1}{3}$
		$\frac{5}{6}$	$.83\frac{1}{3}$			$\frac{11}{12}$	$.91\frac{2}{3}$

Tables for other decimal equivalents such as 16ths, 32nds, and 64ths are available or can be compiled. The 9ths are easy to remember since $\frac{1}{9} = .11\frac{1}{9}$, $\frac{2}{9} = .22\frac{2}{9}$, $\frac{3}{9} = .33\frac{3}{9}$, and so on.

56. Any number that can be divided evenly into another number is an

_________________ part of that number.

- - - - - - - - - - - - - - - -

aliquot

57. $16\frac{2}{3}$ is an aliquot part of 1 because 1 can be divided evenly by $.16\frac{2}{3}$

_____ times.

- - - - - - - - - - - - - - - -

6

58. $.12\frac{1}{2}$, .50, and .20 are all ________________ parts of _____.

- - - - - - - - - - - - - - - -

aliquot; 1

59. Underscore the aliquot parts of $1.00 in the following list:

$0.25, $0.17, 0.33\frac{1}{3}$, $0.20, $0.23

- - - - - - - - - - - - - - - -

$0.25; 0.33\frac{1}{3}$; $0.20

60. List the decimal equivalents of 9ths to 4 decimal places.

__________, __________, __________, __________, __________,

__________, __________, __________

- - - - - - - - - - - - - - - -

.1111, .2222, .3333, .4444, .5556, .6667, .7778, .8889

Using Aliquot Parts of $1.00 in Multiplication

61. If the decimal equivalents are known for the fractions being used, a great deal of time can be saved in many applications, particularly in invoice work. When values such as $.33\frac{1}{3}$ or $.16\frac{2}{3}$ are used, it is easier to divide by 3 or 6 than to multiply by the full decimal. If $.66\frac{2}{3}$ is used, divide by 3 and multiply by 2, since this is equal to $\frac{2}{3}$.

Examples: $108.48 \times .08\frac{1}{3} = 108.48 \times \frac{1}{12} = 9.04$

$57 \times .22\frac{2}{9} = 57 \times \frac{2}{9} = 6\frac{1}{3} \times 2 = 12\frac{2}{3}$

Try these, using aliquot parts:

(a) $32 \times .375 =$

(b) $945 x $.66\frac{2}{3}$

(c) $10.55 \times .20 =$

(d) $13.20 \times .58\frac{1}{3} =$

(e) $2.10 \times .33\frac{1}{3} =$

(f) $144 \times .08\frac{1}{3} =$

(g) $1.25 \times 80 =$

(h) $75.65 \times .60 =$

(i) $10.80 \times .22\frac{2}{9} =$ (j) $84 \times .75 =$

- - - - - - - - - - - - - - -

(a) $32 \times \frac{3}{8} = 12$; (b) $\$945 \times \frac{2}{3} = 630$; (c) $10.55 \times \frac{1}{5} = 2.11$;

(d) $13.20 \times \frac{7}{12} = 7.70$; (e) $2.10 \times \frac{1}{3} = .70$; (f) $144 \times \frac{1}{12} = 12$;

(g) $1.25 \times 80 = 125 \times .80 = 125 \times \frac{4}{5} = 100$; (h) $75.65 \times \frac{3}{5} = 45.39$;

(i) $10.80 \times \frac{2}{9} = 2.40$; (j) $84 \times \frac{3}{4} = 63$

62. Complete the following using aliquot parts:

Example: 39 yd at $\$0.33\frac{1}{3}$ a yd

Solution: $39 \times \frac{1}{3} = \$13.00$ since $.33\frac{1}{3} = \frac{1}{3}$ of $1.00

(a) 95 articles at $12\frac{1}{2}$¢ each =

(b) 108 pieces at $\$0.41\frac{2}{3}$ each =

(c) 32 yd at $37\frac{1}{2}$¢ a yd =

(d) 85 lb at 60¢ a lb =

(e) 120 yd at $\$0.37\frac{1}{2} =$

- - - - - - - - - - - - - - -

(a) $95 \times \frac{1}{8} = 11.875 = \$11.88$; (b) $108 \times \frac{5}{12} = \$45.00$;

(c) $32 \times \frac{3}{8} = \$12.00$; (d) $85 \times \frac{3}{5} = \$51.00$; (e) $120 \times \frac{3}{8} = \$45.00$

SELF-TEST

Before you go to the next chapter take this Self-Test. Compare your answers with those given at the end of the test.

1. Addition

(a) $7\frac{1}{6} + 12\frac{1}{3} + 8\frac{5}{6} =$ ______ (b) $36\frac{1}{3} + 3\frac{2}{3} + 27\frac{1}{4} =$ ______

(c) $6 + 2\frac{3}{4} + 5\frac{3}{9} + 28\frac{3}{5} =$ ______ (d) $2\frac{5}{8} + 5\frac{1}{14} + 7\frac{3}{5} + 16\frac{1}{12} =$ ______

(e) $36\frac{1}{2} + 18\frac{3}{4} + 11\frac{2}{5} + 21\frac{1}{3} =$ ______

2. Subtraction

(a) $28\frac{2}{3} - 8\frac{1}{6} =$ ______ (b) $42\frac{7}{8} - 13\frac{3}{8} =$ ______

(c) $14\frac{1}{9} - 3\frac{2}{9} =$ ______ (d) $24\frac{1}{6} - 21\frac{3}{4} =$ ______

(e) $120 - 101\frac{1}{6} =$ ______

3. Multiplication

(a) $20\frac{1}{2} \times 15\frac{1}{2} =$

(b) $66 \times 4\frac{1}{3} =$

(c) $7\frac{1}{3} \times 81\frac{1}{2} =$

(d) $8\frac{1}{6} \times 5 =$

(e) $1\frac{2}{3} \times 4\frac{1}{4} =$

4. Division

(a) $95 \div 6\frac{1}{3} =$

(b) $28 \div \frac{2}{7} =$

(c) $12\frac{3}{4} \div 3 =$

(d) $14\frac{1}{2} \div 3\frac{3}{4} =$

(e) $\frac{60\frac{1}{2}}{16} =$

5. Change to the decimal form. Record to 2 decimals and carry fractions.
Example: $\frac{2}{9} = .22\frac{2}{9}$

(a) $1\frac{3}{8} =$

(b) $\frac{11}{9} =$

(c) $\frac{1}{12} =$

(d) $\frac{5}{11} =$

(e) $\frac{3}{16} =$

(f) $2\frac{5}{12} =$

(g) $15\frac{1}{6} =$

(h) $9\frac{1}{5} =$

(i) $6\frac{5}{8} =$

(j) $10\frac{3}{4} =$

6. Change to the fractional form.

   Example: $3.91\frac{2}{3} = 3\frac{11}{12}$

   (a) $.16\frac{2}{3} =$ (b) $.222\frac{2}{9} =$

   (c) $1.875 =$ (d) $.0625 =$

   (e) $.9025 =$ (f) $.44\frac{4}{9} =$

   (g) $3.625 =$ (h) $20.83\frac{1}{3} =$

   (i) $4.1875 =$ (j) $.4375 =$

7. Complete the following:

   (a) $48 \times .62\frac{1}{2} =$ (b) $75 \times .33\frac{1}{3} =$

   (c) $108 \times .08\frac{1}{3} =$ (d) $48 \times .16\frac{2}{3} =$

   (e) $36 \times .22\frac{2}{9} =$ (f) $96 \times .87\frac{1}{2} =$

   (g) $45 \times .60 =$ (h) $120 \times .75 =$

   (i) $132 \times .75 =$ (j) $48 \times .58\frac{1}{3} =$

8. Find the cost of the following purchases. All prices are per unit.

   (a) 36 lb at $66\frac{2}{3}$¢ = (b) 90 lb at 20¢ =

   (c) 72 lb at $37\frac{1}{2}$¢ = (d) 24 ft at $83\frac{1}{3}$¢ =

   (e) 80 ft at $\$0.12\frac{1}{2}$ = (f) 60 qt at $\$0.41\frac{2}{3}$ =

   (g) 54 qt at $\$0.11\frac{1}{2}$ = (h) 24 qt at $\$0.16\frac{2}{3}$ =

   (i) 35 pc at \$0.80 = (j) 160 pc at \$0.75 =

Answers to Self-Test

Compare your answers to the Self-Test with the answers given here. If you miss any questions, review the frames indicated in parentheses before continuing to the next chapter. If you think you need more practice, additional problems are offered in the Appendix.

1. (a)

$$\begin{array}{r|l} 7\frac{1}{6} & \frac{1}{6} \\ 12\frac{1}{3} & \frac{2}{6} \\ 8\frac{5}{6} & \frac{5}{6} \\ \scriptstyle 1 & \\ \hline 28\frac{1}{3} & \frac{8}{6} = 1\frac{2}{6} = 1\frac{1}{3} \end{array}$$

(b)

$$\begin{array}{r|l} 36\frac{1}{3} & \frac{4}{12} \\ 3\frac{2}{3} & \frac{8}{12} \\ 27\frac{1}{4} & \frac{3}{12} \\ \scriptstyle 1 & \\ \hline 67\frac{1}{4} & \frac{15}{12} = 1\frac{3}{12} = 1\frac{1}{4} \end{array}$$

(c)

$$\begin{array}{r|l} 6 & \\ 2\frac{3}{4} & \frac{135}{180} \\ 5\frac{3}{9} & \frac{60}{180} \\ 28\frac{3}{5} & \frac{108}{180} \\ \scriptstyle 1 & \\ \hline 42\frac{41}{60} & 1\frac{123}{180} = 1\frac{41}{60} \end{array}$$

or by reducing $5\frac{3}{9}$ to $5\frac{1}{3}$, the problem is easier to add, i.e.

$$\begin{array}{r|l} 6 & \\ 2\frac{3}{4} & \frac{45}{60} \\ 5\frac{1}{3} & \frac{20}{60} \\ 28\frac{3}{5} & \frac{36}{60} \\ \scriptstyle 1 & \\ \hline 42\frac{41}{60} & \frac{101}{60} = 1\frac{41}{60} \end{array}$$

(d)

$$\begin{array}{r|l} 2\frac{5}{8} & \frac{525}{840} \\ 5\frac{1}{14} & \frac{60}{840} \\ 7\frac{3}{5} & \frac{504}{840} \\ 16\frac{1}{12} & \frac{70}{840} \\ \scriptstyle 1 & \\ \hline 31\frac{319}{840} & \frac{1159}{840} = 1\frac{319}{840} \end{array}$$

$$\begin{array}{r|rrrr} 4 & 8 & 14 & 5 & 12 \\ \hline 2 & 2 & 14 & 5 & 3 \\ \hline & 1 & 7 & 5 & 3 \end{array}$$

LCD = 4 x 2 x 7 x 5 x 3 = 840

(e)

$$\begin{array}{r|l} 36\frac{1}{2} & \frac{30}{60} \\ 18\frac{3}{4} & \frac{45}{60} \\ 11\frac{2}{5} & \frac{24}{60} \\ 21\frac{1}{3} & \frac{20}{60} \\ \scriptstyle 1 & \\ \hline 87\frac{59}{60} & \frac{119}{60} = 1\frac{59}{60} \end{array}$$

(Frames 18, 21-27)

2. (a)

$28\frac{2}{3}$	$\frac{4}{6}$
$-\ 8\frac{1}{6}$	$\frac{1}{6}$
$20\frac{1}{2}$	$\frac{3}{6} = \frac{1}{2}$

(b)

$$\begin{array}{r} 42\frac{7}{8} \\ -\ 13\frac{3}{8} \\ \hline 29\frac{4}{8} = 29\frac{1}{2} \end{array}$$

(c)

$$\begin{array}{r} 13\frac{10}{9} \\ \not{1}\not{4}\frac{\not{1}}{\not{9}} \\ -\ 3\frac{2}{9} \\ \hline 10\frac{8}{9} \end{array}$$

(d)

23	$\frac{14}{12}$
$\not{2}\not{4}\frac{1}{6}$	$\frac{\not{2}}{\not{1}\not{2}}$
$-\ 21\frac{3}{4}$	$\frac{9}{12}$
$2\frac{5}{12}$	$\frac{5}{12}$

(e)

$$\begin{array}{r} 119\frac{6}{6} \\ \not{1}\not{2}\not{0} \\ -\ 101\frac{1}{6} \\ \hline 18\frac{5}{6} \end{array}$$

(Frames 30, 33)

3. (a) $20\frac{1}{2} \times 15\frac{1}{2} = \frac{41}{2} \times \frac{31}{2} = \frac{1271}{4} = 317\frac{3}{4}$

(b) $66 \times 4\frac{1}{3} = \overset{22}{\not{6}\not{6}} \times \frac{13}{\not{3}} = 286$

(c) $7\frac{1}{3} \times 81\frac{1}{2} = \frac{\overset{11}{\not{2}\not{2}}}{3} \times \frac{163}{\not{2}} = \frac{1793}{3} = 597\frac{2}{3}$

(d) $8\frac{1}{6} \times 5 = \frac{49}{6} \times 5 = \frac{245}{6} = 40\frac{5}{6}$

(e) $1\frac{2}{3} \times 4\frac{1}{4} = \frac{5}{3} \times \frac{17}{4} = \frac{85}{12} = 7\frac{1}{12}$ (Frames 34-38)

4. (a) $95 \div 6\frac{1}{3} = 95 \div \frac{19}{3} = \overset{5}{\not{9}\not{5}} \times \frac{3}{\not{1}\not{9}} = 15$

(b) $28 \div \frac{2}{7} = \overset{14}{\not{2}\not{8}} \times \frac{7}{\not{2}} = 98$

(c) $12\frac{3}{4} \div 3 = \frac{\overset{17}{\not{5}\not{1}}}{4} \times \frac{1}{\not{3}} = \frac{17}{4} = 4\frac{1}{4}$

(d) $14\frac{1}{2} \div 3\frac{3}{4} = \frac{29}{2} \div \frac{15}{4} = \frac{29}{\not{2}} \times \frac{\overset{2}{\not{4}}}{15} = \frac{58}{15} = 3\frac{13}{15}$

(e) $\frac{60\frac{1}{2}}{16} = \frac{60\frac{1}{2} \times 2}{16 \times 2} = \frac{121}{32} = 3\frac{25}{32}$ (using a common multiple)

or $60\frac{1}{2} \div 16 = \frac{121}{2} \times \frac{1}{16} = \frac{121}{32} = 3\frac{25}{32}$ (Frames 40-42)

5. (a) $1.37\frac{1}{2}$; (b) $1.22\frac{2}{9}$; (c) $.08\frac{1}{3}$; (d) $.45\frac{5}{11}$; (e) $.18\frac{3}{4}$; (f) $2.41\frac{2}{3}$

(g) $15.16\frac{2}{3}$; (h) 9.20; (i) $6.62\frac{1}{2}$; (j) 10.75 (Frame 43)

6. (a) $\frac{1}{6}$; (b) $\frac{2}{9}$; (c) $1\frac{7}{8}$; (d) $\frac{1}{16}$; (e) $\frac{361}{400}$; (f) $\frac{4}{9}$; (g) $3\frac{5}{8}$; (h) $20\frac{5}{6}$;

(i) $4\frac{3}{16}$; (j) $\frac{7}{16}$ (Frame 43)

7. (a) $48 \times \frac{5}{8} = \$30.00$; (b) $75 \times \frac{1}{3} = \$25.00$; (c) $108 \times \frac{1}{12} = \$9.00$;

(d) $48 \times \frac{1}{6} = \$8.00$; (e) $36 \times \frac{2}{9} = \$8.00$; (f) $96 \times \frac{7}{8} = \$84.00$;

(g) $45 \times \frac{3}{5} = \$27.00$; (h) $120 \times \frac{3}{4} = \$90.00$; (i) $132 \times \frac{3}{4} = \$99.00$;

(j) $48 \times \frac{7}{12} = \$28.00$ (Frames 55, 61-62)

8. (a) $36 \times \frac{2}{3} = \$24.00$; (b) $90 \times \frac{1}{5} = \$18.00$; (c) $72 \times \frac{3}{8} = \$27.00$;

(d) $24 \times \frac{5}{6} = \$20.00$; (e) $80 \times \frac{1}{8} = \$10.00$; (f) $60 \times \frac{5}{12} = \$25.00$;

(g) $54 \times \frac{1}{9} = \$6.00$; (h) $24 \times \frac{1}{6} = \$4.00$; (i) $35 \times \frac{4}{5} = \$28.00$;

(j) $160 \times \frac{3}{4} = \$120.00$ (Frames 61-62)

CHAPTER FOUR

Percentage

OBJECTIVES

In this chapter you will learn how to

- change any whole number, mixed number, decimal, fraction, or percent to any of the other forms;
- use aliquot parts instead of a percent, when possible, in multiplication work;
- find the percentage (amount) when the rate and base are known;
- find the rate (percent) when the percentage and base are known;
- find the base (100%) when the percentage and rate are known.

Unit 1

CHANGING PERCENTS TO DECIMALS OR FRACTIONS AND CONVERSELY

1. Try the following problems to see if you can skip this unit.

(a) Express the following percents in the decimal form:

$12\% =$	$21\frac{1}{2}\% =$
$\frac{1}{2}\% =$	$125.6\% =$

(b) Express the following decimals in the percent form:

$.06\frac{1}{3} =$	$.0045 =$
$1.05 =$	$.165 =$

(c) Express the following percents in the fractional form:

$18\% =$ $\qquad$ $16\frac{1}{2}\% =$

$37\frac{1}{2}\% =$ $\qquad$ $210\% =$

(d) Express the following fractions in the decimal form and the percent form:

$\frac{3}{8} = .375 = 37\frac{1}{2}\%$ $\qquad$ $\frac{3}{12} =$

$\frac{5}{6} =$ $\qquad$ $\frac{12}{92} =$

(e) Express each of the following percents as a decimal and as a fraction or mixed number:

$25\% = .25 = \frac{1}{4}$ $\qquad$ $\frac{1}{2}\% =$

$16\% =$ $\qquad$ $1\frac{1}{4}\% =$

(f) Show the decimal forms of the following. Note the relationships between them.

$\frac{1}{2}$ and $\frac{1}{2}\%$? $\frac{1}{2} = .50$, $\frac{1}{2}\% = .005$

$\frac{1}{4}$ and $\frac{1}{4}\%$? $\frac{1}{4} =$ $\qquad$ $\frac{1}{4}\% =$

$\frac{1}{8}$ and $\frac{1}{8}\%$? $\frac{1}{8} =$ $\qquad$ $\frac{1}{8}\% =$

(a) .12, .215, .005, 1.256; (b) $6\frac{1}{3}\%$, .45%, 105%, 16.5%; (c) $\frac{9}{50}$, $\frac{33}{200}$, $\frac{3}{8}$, $2\frac{1}{10}$; (d) .25 = 25%, $.83\frac{1}{3} = 83\frac{1}{3}\%$, .1304 = 13.04%; (e) $.005 = \frac{1}{200}$, $.16 = \frac{4}{25}$, $.0125 = \frac{1}{80}$; (f) .25, .0025; .125, .00125

If you had any difficulty or made any errors, go to the next frame. Otherwise skip ahead to Frame 12.

2. In the preceding chapter you learned that it was easier to add, subtract, and compare fractions with the same denominator. For example, it is easier to add $\frac{34}{100}$ and $\frac{25}{100}$ than to add $\frac{17}{50}$ and $\frac{1}{4}$, yet their sum is the same.

   Percent is a special kind of fraction because the denominator is always 100. Percent means hundredths:

   $\frac{12}{100}$ = 12 percent, $\frac{16\frac{1}{2}}{100} = 16\frac{1}{2}$ percent, $\frac{200}{100}$ = 200 percent

   Then, $\frac{15}{100}$ = _____ percent, $\frac{2\frac{1}{2}}{100}$ = _____ percent, and $\frac{125}{100}$ = _____ percent.

   - - - - - - - - - - - - - - - -

   15, $2\frac{1}{2}$, 125

3. The fractional expression "hundredths" came to be replaced with the symbol % which means percent. Thus,

   one hundredth ($\frac{1}{100}$) of a quantity may be expressed as 1%,

   $\frac{2}{100}$ as 2%, $\frac{3}{100}$ as 3%, and so forth.

   4 percent is the same as $\frac{4}{100}$ or ______%

   6 percent is the same as _______ or ______%

   5 percent is the same as _______ or ______%

   - - - - - - - - - - - - - - - -

   4%, $\frac{6}{100}$ or 6%, $\frac{5}{100}$ or 5%

4. Since hundredths may be written in the decimal form, percents may be expressed in three ways:

   $1\% = \frac{1}{100} = .01$, $3\% = \frac{3}{100} = .03$, $20\% = \frac{20}{100} = .20$, and so on.

   14 percent may be written as _______ (percent), _______ (fraction), or _______ (decimal).

   - - - - - - - - - - - - - - - -

   14%, $\frac{14}{100}$, or .14

5. The percent sign takes the place of the two decimal places that denote hundredths:

$$1\% = \frac{1}{100} = .01; \text{ conversely, } .01 = \frac{1}{100} = 1\%$$

To change a percent to a decimal drop the percent sign and move the decimal point two places to the left. Then 4% = .04; 125% = 1.25; 2.5% = .025; $\frac{1}{2}\%$ = .005. The following percents written as decimals are

12% = __________, $3\frac{1}{2}\%$ = __________, 106% = __________

- - - - - - - - - - - - - - - -

.12, .035, 1.06

6. To change a decimal to a percent move the decimal point two places to the right and use the percent sign. Then .16 = 16%; .045 = 4.5%; $.12\frac{1}{2} = 12\frac{1}{2}\%$ or 12.5%; .0056 = .56%. The following decimals written as percents are

.25 = ______%, $.20\frac{1}{2}$ = ______%, 2.26 = ______%, .0025 = ______%

- - - - - - - - - - - - - - - -

25%, $20\frac{1}{2}\%$, 226%, .25%

7. Since percent denotes hundredths, all percents may be thought of automatically as fractions with a denominator of 100. To change a percent to a fraction drop the percent sign, divide the percent quantity by 100, and reduce to lowest terms. Then,

$$60\% = \frac{60}{100} = \frac{3}{5}, \quad 4\% = \frac{4}{100} = \frac{1}{25}, \quad 37\tfrac{1}{2}\% = \frac{37\frac{1}{2}}{100} = \frac{3}{8}$$

What are the fractions for the following:

40% = __________, 23% = __________, 12% = __________

- - - - - - - - - - - - - - - -

$\frac{40}{100} = \frac{2}{5}, \frac{23}{100}, \frac{12}{100} = \frac{3}{25}$

8. If $16\frac{2}{3}\%$ were written as a fraction with a denominator of 100, it would be $\frac{16\frac{2}{3}}{100}$. In order to reduce it to a common fraction, study the following procedure.

(a) $\frac{16\frac{2}{3}}{100} = 16\frac{2}{3} \div 100 = \frac{50}{3} \times \frac{1}{100} = \frac{1}{6}$

or (b) $\frac{16\frac{2}{3}}{100} = \frac{16\frac{2}{3} \times 3}{100 \times 3} = \frac{50}{300} = \frac{1}{6}$

Change $22\frac{2}{9}\%$ to a fraction and reduce to lowest terms.

$22\frac{2}{9} \div 100 = \frac{200}{9} \times \frac{1}{100} = \frac{200}{900} = \frac{2}{9}$, or $\frac{22\frac{2}{9} \times 9}{100 \times 9} = \frac{200}{900} = \frac{2}{9}$

9. If $12\% = \frac{12}{100} = \frac{3}{25}$, then the reverse is true. <u>To change a fraction to a percent</u>, divide the numerator by the denominator to obtain the decimal form and then change the decimal form to a percent. Carry answers for problems (d) through (g) to <u>four</u> decimals and correct to <u>three</u>.

Example: $\frac{3}{8} = .375 = 37.5\%$ or $37\frac{1}{2}\%$

(a) $\frac{5}{8} = .625 =$ (b) $\frac{11}{5} = 2.2 =$

(c) $\frac{3}{5} =$ (d) $\frac{23}{56} = .411 =$

(e) $\frac{4}{9} = .444 =$ (f) $\frac{6}{25} =$

(g) $\frac{32}{57} =$

(a) 62.5% or $62\frac{1}{2}\%$; (b) 220%; (c) .60 = 60%; (d) 41.1%;
(e) 44.4% (or could be recorded as $44\frac{4}{9}\%$); (f) .240 = 24.0%;
(g) .561 = 56.1%

10. Any whole number such as 1 or 2 may be written as 1.00, 2.00. Then 1 = 1.00 = 100%, 2 = 2.00 = 200%, and so forth. If .25 = 25% and 1.00 = 100%, then 1.25 = 125%. Express the following numbers in the percent form:

3 = _______%, 3.5 = _______%, 4.25 = _______%, 20 = _______%

300%, 350%, 425%, 2000%

11. A mixed number such as $2\frac{1}{2}$, when written as a decimal, is 2.5 or 2.50, which is 250%. Study the following:

$$6\frac{1}{8} = 6.125 = 612.5\% \text{ or } 612\frac{1}{2}\%; \quad 4\frac{1}{3} = 4.33\frac{1}{3} = 433\frac{1}{3}\%$$

Then $1\frac{2}{5}$ = __________ (decimal) = __________ (percent)

$3\frac{1}{2}$ = __________ (decimal) = __________ (percent)

$4\frac{2}{3}$ = __________ (decimal) = __________ (percent)

- - - - - - - - - - - - - - - -

$1.40 = 140\%; \; 3.50 = 350\%; \; 4.66\frac{2}{3} = 466\frac{2}{3}\%$

Aliquot Parts

12. When percents are used to solve a problem, the decimal form is generally used. However, it is sometimes more convenient to use the fractional form, as shown in the following examples.

Example: $8\frac{1}{3}\%$ of $144 = .083333 \times 144 = 11.9995$ or 12

Procedure: This problem is a simple one if we know that $8\frac{1}{3}\% = \frac{1}{12}$, in which case $144 \div 12 = 12$.

Example: $12\frac{1}{2}\%$ of $96 = \frac{1}{8}$ of $96 = 12$

Procedure: This is less cumbersome than multiplying 96 by .125 (decimal equivalent of $12\frac{1}{2}\%$).

We discussed aliquot parts when we were multiplying with fractions (Chapter 3). Now that we can change a percent to a fraction, we can use this same information (and the table on page 90) when we want to use a percent instead of a fraction.

13. If 25% of 324 is the same as $\frac{1}{4} \times 324 = 81$, then 50% of 28 is the same as __________ x 28 = ________, and $16\frac{2}{3}\%$ of 360 is the same as __________ x 360 = ________.

- - - - - - - - - - - - - - - -

$\frac{1}{2} \times 28 = 14, \; \frac{1}{6} \times 360 = 60$

14. How would you figure out that $41\frac{2}{3}\% = \frac{5}{12}$? If you didn't remember this fact and didn't have a table of reference, you would have to divide $41\frac{2}{3}$ by ________ and __________ to its lowest terms.

- - - - - - - - - - - - - - - -

100, reduce

15. Here is the procedure: $41\frac{2}{3}\% = \frac{41\frac{2}{3}}{100} = \frac{125}{3} \times \frac{1}{100} = \frac{5}{12}$

Now change $87\frac{1}{2}\%$ to a common fraction.

- - - - - - - - - - - - - - - -

$87\frac{1}{2}\% = \frac{87\frac{1}{2}}{100} = \frac{175}{2} \times \frac{1}{100} = \frac{7}{8}$

Unit 2

BASIC ELEMENTS OF PERCENT PROBLEMS

16. Before beginning this unit see if you can solve the following problems.

(a) State the basic percentage formula ____________________

(b) (1) In the statement 15% of 82 = 12.30, the rate is _______, the percentage is _______, and the base is _______.

(2) In any percentage problem, the base is always _______%.

(c) Find the answers to the following correct to two decimals.

(1) What is 23% of 96? ____________________

2.6% of 300? ____________________

$\frac{1}{2}\%$ of \$150? ____________________

.08% of 21.5? ____________________

$112\frac{1}{2}\%$ of 144? ____________________

(2) What is 26 increased by 12%? ____________________

(3) What is 125 decreased by 15%? ____________________

(d) If last year's school attendance of 3700 increased by 12% this year, what would this year's attendance equal?

(e) If Henry's earnings of $75.00 per week were reduced 6%, what would his new earnings be?

- - - - - - - - - - - - - - -

(a) base x rate = percentage
(b) (1) rate is 15%, percentage is 12.30, base is 82; (2) 100%
(c) (1) .23 x 96 = 22.08; .026 x 300 = 7.80; .005 x $150 = $0.75; .0008 x 21.5 = .01720 = .02; 1.125 x 144 = 162
(2) 26 x 1.12 = 29.12; (3) 125 x .85 = 106.25
(d) 3700 x 1.12 = 4144; (e) $75 x .94 = $70.50

If you had no difficulty or made no errors, skip to Frame 25. Otherwise continue on here.

17. There are only three basic factors or elements in any percentage problem. These are the base (100% or the whole amount of anything), the rate (percent), and the percentage (amount). The relationships of these factors are expressed as a formula,

Base x Rate = Percentage

by which all percentage problems can be solved. The secret to solving a problem successfully is the ability to identify the factors that are known and the factor that is to be found.

The three basic elements are the __________, __________, and ____________________.

- - - - - - - - - - - - - - -

base, rate, and percentage

18. If 15% of 32 = .15 x 32 = 4.80, the rate is 15%, the base is 32, and the percentage is 4.80.

In the problem 45% of 250 = 112.50, the rate is ________, the base is ________ and the percentage is __________.

- - - - - - - - - - - - - - -

45%, 250, 112.50

19. The base always represents the whole amount of anything or

_______%.

- - - - - - - - - - - - - - - -

100%

20. If 12% of 620 = 74.40, the base is _______ and the rate is _______.

- - - - - - - - - - - - - - - -

620, 12%

<u>Finding the Percentage (Amount)</u>

<u>Base x Rate = Percentage (or B x R = P)</u>

21. Let's look at some examples that show how to find the percentage.

<u>Example</u>: Find 12% of 125.

<u>Solution</u>: 125 x .12 = 15

What is wanted here is 12 parts (that is 12 hundredths) of 125 or $\frac{12}{100}$ x 125. The decimal form is generally used, although sometimes the fractional form is easier to use arithmetically.

<u>Example</u>: Mr. Smith earns $7900 a year. He saves 8% of this amount. How much does he save?

<u>Solution</u>: $7900 x .08 = $632 (amount saved yearly)

In this example, the base is _______, the rate is _______, and the percentage is _______.

- - - - - - - - - - - - - - - -

$7900, 8%, $632

22. To find the percentage (amount), all rates expressed as a percent must be changed to the decimal or fractional form before multiplying by any other number.

(a) 32% of 65 = .32 x 65 = __________
(rate) (base) (rate) (base) (percentage)

(b) 25% of 220 = ________________ x ______ = __________
(rate as a decimal) (base) (percentage)

or ________________ x ______ = __________
(rate as a fraction) (base) (percentage)

(c) 110% of 456 = 1.10 x 456 = __________

- - - - - - - - - - - - - - - -

(a) 20.80; (b) 25% of 220 = .25 x 220 = 55 or $\frac{1}{4}$ x 220 = 55;
(c) 501.6

23. Now try working these problems:

(a) 4% of 86 ________ x ________ = ________

(b) 130% of 540 ________ x ________ = ________

(c) 125½% of 62 ________ x ________ = ________

- - - - - - - - - - - - - - - -

(a) .04 x 86 = 3.44; (b) 1.30 x 540 = 702.00; (c) 1.255 x 62 = 77.810

24. Find the percentage if the base is $1200 and the rate is 20%.

________ x ________ = ________

- - - - - - - - - - - - - - - -

.20 or $\frac{1}{5}$ x 1200 = $240

Shortcuts in Finding the Percentage for Certain Rates

25. In addition to the use of aliquot parts, there are some additional shortcuts that may be used to advantage in obtaining percentages for certain rates.

Example: 10% of $136 = $13.60

Solution: Since 10% = .10 = $\frac{1}{10}$, we can divide by 10 to obtain 10% of any number. We already learned that to divide by 10, we move the decimal one place to the left. Thus,

10% of 8.26 = .826; 10% of 31.8 = 3.18, etc.

What is 10% of 1.85? ______ of .96? ______ of $21.00? ______

- - - - - - - - - - - - - - - -

.185; .096; $2.10

26. If we can readily find 10%, then there is an easy and quick way to find 5% of some number other than multiplying by 5.

Example: Find 5% of $126.40

Solution: (1) From Frame 25, we know that 10% = $12.64.
(2) Then 5% = $\frac{1}{2}$ of 10% = $6.32.

With practice, step (1) is unnecessary (i.e., mentally move the decimal point to get 10% and then divide by 2). Try these:

(a) 5% of 32.50 = ________ (b) 5% of .86 = ________

- - - - - - - - - - - - - - - -

(a) 1.625; (b) .043

27. From the last two frames, we see that we can also obtain 15% quickly by adding 10% and 5%, instead of multiplying by .15.

Example: Find 15% of $96.80

Solution:
10% = $ 9.68
5% = 4.84
15% = $14.52

Example: Find 15% of 37.29

Solution:
10% = 3.729
5% = 1.8645
15% = 5.5935

If 37.29 = $37.29, the 15% would be recorded as $5.59.

Find 15% of

(a) 840 (b) 937 (c) 2.06

- - - - - - - - - - - - - - - -

(a) 84.0
42.0
126.0

(b) 93.70
46.85
140.55

(c) .206
.103
.309

28. To find $2\frac{1}{2}$% of any number, take 10% and divide by 4.

Example: What is $2\frac{1}{2}$% of 382?

Solution: 10% = 38.2
$2\frac{1}{2}\% = 38.2 \div 4 = 9.55$

What is $2\frac{1}{2}$% of

(a) 420?

(b) 16.92?

- - - - - - - - - - - - - - - -

(a) $10\% = 42$, $2\frac{1}{2}\% = 42 \div 4 = 10.5$;
(b) $10\% = 1.692$, $2\frac{1}{2}\% = 1.692 \div 4 = .423$

29. We have seen how easy it is to obtain 10%, 5%, 15%, and $2\frac{1}{2}\%$. Now suppose we want to find 85% of some number. This is the same as taking 15% from that number.

Example: Find 85% of 126 $(.85 \times 126 = 107.1)$

Solution: 15% of $126 = 12.6 + 6.3 = 18.9$
$126 - 18.9 = 107.1$

This is much easier with less chance of error than multiplying by .85.

Try this (using shortcut method):

85% of 13.2 =

- - - - - - - - - - - - - - - -

$15\% = 1.32 + .66 = 1.98$, $85\% = 13.2 - 1.98 = 11.22$

30. In the same way, we can obtain 95% by subtracting _______% from the number, 110% by adding _______% to the number, and $97\frac{1}{2}\%$ by subtracting _______ from the number.

- - - - - - - - - - - - - - - -

5%, 10%, $2\frac{1}{2}\%$

31. Using shortcut methods, find

(a) 95% of \$820

(b) 110% of 960

(c) $97\frac{1}{2}\%$ of 26

- - - - - - - - - - - - - - - -

(a) $5\% = 41$, $95\% = 820 - 41 = \$779$
(b) $10\% = 96$, $110\% = 960 + 96 = 1056$
(c) $2\frac{1}{2}\% = \frac{1}{4} \times 2.60 = .65$, $97\frac{1}{2}\% = 26 - .65 = 25.35$

32. In the examples below we show a quick way of finding a fractional part of 1% such as $\frac{1}{2}$%, or $\frac{1}{4}$%.

Example: $\frac{1}{2}$% of 28 — Solution: 1% = .28 (divide by 100 since 1% = $\frac{1}{100}$), then $\frac{1}{2}$% = $\frac{1}{2}$ of .28 = .14

Example: $\frac{1}{4}$% of 28 — Solution: $\frac{1}{4}$% = $\frac{1}{4}$ of .28 = .07

Any fraction of 1% may be obtained in the same way. Try these:

(a) $\frac{1}{5}$% of 820 (b) $\frac{1}{3}$% of 96

(a) 1% = 8.20
$\frac{1}{5}$% = 8.20 ÷ 5 = 1.64

(b) 1% = .96
$\frac{1}{3}$% = .96 ÷ 3 = .32

33. Find the percentage for the following problems (use shortcuts when possible).

(a) $12\frac{1}{2}$% of \$360 =

(b) $\frac{1}{4}$% of 96 =

(c) 113% of \$240 =

(d) 85% of 83 =

(a) .125 x \$360 = \$45 or $\frac{1}{8}$ x \$360 = \$45

(b) .0025 x 96 = .24 or 1% = .96, then $\frac{1}{4}$% = $\frac{1}{4}$ of .96 = .24

(c) 1.13 x 240 = 271.20

(d) .15 x 83 = 8.30 + 4.15 = 12.45; 85% = 83 - 12.45 = 70.55

34. How good are you at finding the rate? Try these problems to see if you can skip the next section.

(a) What percent is 15 of 45?

27 of 86?

152 of 64?

5.43 of 36.2?

$\frac{5}{6}$ of $\frac{2}{3}$?

(b) If the student enrollment of Haven's School were 3462 last year and 4085 this year, what percent increase does this represent?

(c) Sales for Atlin Corp. dropped from \$36,000,000 to \$26,000,000. What percent decrease does this represent?

(a) $15 \div 45 = .33\frac{1}{3} = 33\frac{1}{3}\%$

$27 \div 86 = .31395 = 31.40\%$ or $.31\frac{34}{86} = .31\frac{17}{43} = 31\frac{17}{43}\%$

$152 \div 64 = 2.375 = 237.5\%$

$5.43 \div 36.2 = .15 = 15\%$

$\frac{5}{6} \div \frac{2}{3} = 1.25 = 125\%$

(b) $4085 \div 3462 = 1.179 = 1.18 = 118\%$

Increase $= 118\% - 100\% = 18\%$

(c) $\$26,000,000 \div \$36,000,000 = 26 \div 36 = 72\frac{2}{9}\%$

Decrease $= 100\% - 72\frac{2}{9}\% = 27\frac{7}{9}\%$ (28% would be acceptable)

If you had no difficulty or made no mistakes, skip to Frame 40. Otherwise continue on here.

Finding the Rate (Percent)

Rate = Percentage ÷ Base (R = P ÷ B)

35. We will start with some examples for finding the rate.

Example: 30 is what percent of 180?

Solution: $30 \div 180 = \frac{1}{6} = .16\frac{2}{3} = 16\frac{2}{3}\%$

In this case we wish to find out what fractional part 30 is of 180, expressed as a percent. We find that it is $16\frac{2}{3}$ parts (hundredths) or $16\frac{2}{3}\%$.

Example: In a class of 35 students 28 received passing grades. What percent did this represent?

Solution: $28 \div 35 = \frac{4}{5} = .80 = 80\%$

Example: The sales for Benite Company amounted to $36,462 last week and $35,400 the week before. What percent increase does this represent?

Solution: This problem may be solved by two methods.
$36,462 - $35,400 = $1,062 increase in sales
$1062 ÷ $35,400 = .03 or 3% increase
or
$36,462 ÷ $35,400 = 1.03 or 103%. This means that last week's sales amounted to 103% of the preceding week's sales, or an increase of 3%.

36. If 713,400 voters out of 870,000 voted for a certain bond issue, what percent of voters favored the issue? We know the base is 870,000 and the percentage is 713,400, but we don't know what the rate is.

________ ÷ ______ = ________ = ________
(percentage) (base) (decimal rate) (percent rate)

- - - - - - - - - - - - - - -

713,400 ÷ 870,000 = .82 = 82%

37. Out of a class of 105 students 84 received passing grades. What percent does this represent?

(a) The base is ______, the percentage is ______.

(b) Rate is ______ ÷ ______ = ______%.

- - - - - - - - - - - - - - -

(a) 105, 84; (b) 84 ÷ 105 = .80 = 80%

38. It is possible for the rate to be greater than 100%. If last week's sales for a store were $437,000 and this week's sales amounted to $489,440, what percent is this week's sales of last week's sales? What is the percent increase?

(a) In this problem the base is ________________.

(b) ____________ ÷ ____________ = _____%, % increase = ____

Keep the following two thoughts in mind when solving a problem of this sort:

- Whenever you want to find a rate, you are going to divide.
- The word "of" when used, precedes the divisor (base).

- - - - - - - - - - - - - - - -

(a) last week's sales, or \$437,000
(b) \$489,440 ÷ \$437,000 = 1.12 or 112%, increase = 12%

39. Find the rate (to 2 decimal places), given the percentage and base for the following. Show all remainders as a fraction.

Example: 16 is what percent of 98? $16 \div 98 = .16\frac{16}{49}$ or $16\frac{16}{49}\%$

(a) What percent of 96 is 24? ________ ÷ ________ = ________%

(b) 15 is what percent of 300? ________ ÷ ________ = ________%

(c) 43 is what percent of 344? ________ ÷ ________ = ________%

(d) What percent of 16 is 64? ________ ÷ ________ = ________%

- - - - - - - - - - - - - - - -

(a) 24 ÷ 96 = 25%; (b) 15 ÷ 300 = 5%; (c) 43 ÷ 344 = $12\frac{1}{2}\%$;
(d) 64 ÷ 16 = 400%

40. In the next section we will deal with the base or 100%. If you can solve the following problems, you may be able to skip some of the material.

(a) 156.50 is 125% of what number? __________

35 is $33\frac{1}{3}\%$ of what number? __________

$4\frac{1}{2}\%$ of ________ = $67\frac{1}{2}$.

250 is $2\frac{1}{2}\%$ of what number? __________

(b) The \$2700 collected by the local charity drive last year represented 12% of its goal for the year. What is the goal?

(c) Mary Haynes earns \$450 per month this year. This is an increase of 25% over last year. How much did she earn per month last year?

(d) If 40% or 1800 persons in a community are under 18 years of age, what is the total population?

- - - - - - - - - - - - - -

(a) $156.50 \div 1.25 = 125\frac{1}{5}$ or 125.20

$35 \div .33\frac{1}{3}$ or $35 \times 3 = 105$

$67\frac{1}{2} \div .045 = 1500$ or $4\frac{1}{2}\%$ of $1500 = 67\frac{1}{2}$

$250 \div .025 = 10,000$

(b) $\$2700 \div .12 = \$22,500$

(c) $\$450 = 125\%$ of last year, therefore 100% (last year) = $\$450 \div 1.25 = \$360$

(d) $1800 \div .40 = 4500$ or $18 \div \frac{2}{5} = 4500$

If you made any errors or had difficulty, continue here. Otherwise go to the Self-Test at the end of this chapter.

Finding the Base or 100%

Base = Percentage ÷ Rate (B = P ÷ R)

41. Let's look at some examples for finding the base.

Example: 54 is 15% of what number?

Solution: This is the same as saying that 15% (15 hundredths) of some number equals 54. What is that number? If 15% = 54, then

$$1\% = 54 \div 15 = 3.60,\quad 100\% = 3.60 \times 100 = 360$$

These steps can be completed at the same time (i.e., $54 \div .15 = 360$).

Now complete the following statement from this example.

(a) ________ ÷ ____ = ________
percentage rate base (100%)

(b) ____ x ____ = ________
base rate percentage

- - - - - - - - - - - - - -

(a) $54 \div .15 = 360$; (b) $360 \times .15 = 54$

42. Example: If the price of milk is now 22¢ a quart and represents an increase of 10% over the last month's price, what was the cost of milk at that time?

Solution: 22¢ = 110% of last month's price.
Therefore, .22 ÷ 1.10 = .20 or 20¢

Proof: 10% of 20¢ = 2¢ (amount of increase)

In this example, 22¢ = (base/rate/percentage)?
100% + 10% = (base/rate/percentage)?

What factor is missing in the question? ____________

- - - - - - - - - - - - - - - -

percentage, rate, base

43. Example: If 65% or 468 of a school's population are girls, how many boys are there?

Solution: 468 ÷ .65 = 720 school population
720 - 468 = 252 boys

Proof: If 65% were girls, then 100% - 65% or 35% = boys;
35% of 720 = 252.

In this example, 65% = (base/rate/percentage)?
468 = (base/rate/percentage)?

What factor is missing in the question? ____________

- - - - - - - - - - - - - - - -

rate, percentage, base

44. Percentage ÷ rate = base; 26 is 32% of what number? Prove your answer.

26	÷	.32	=	$81\frac{1}{4}$
percentage		rate		base

648 is 120% of what number? (a) __________ ÷ ____ = ____
percentage rate base

(b) Proof: __________________

- - - - - - - - - - - - - - - -

(a) 648 ÷ 1.20 = 540; (b) 540 x 1.20 = 648

45. When we want to find the base, the ____________ is always the divisor. Can the base be less than the percentage? (yes/no)

- - - - - - - - - - - - - - -

rate; yes, when the rate is more than 100%

46. Complete the following:

(a) percentage = ______________ x ______________

(b) rate = ______________ ÷ ______________

(c) base = ______________ ÷ ______________

- - - - - - - - - - - - - - -

(a) base x rate; (b) percentage ÷ base; (c) percentage ÷ rate

47. Solve the following:

(a) Find $6\frac{1}{2}$% of 28.

(b) What percent is 20 of 35?

(c) 45 is 18% of what number?

- - - - - - - - - - - - - - -

(a) .065 x 28 = 1.820; (b) 20 ÷ 35 = $.57\frac{1}{7} = 57\frac{1}{7}$%;
(c) 45 ÷ .18 = 250

48. Mary Smith received an increase in salary from $475 per month last year to $570 this year. What percent increase does this represent? (Since last year's figure was increased, it will have to be the divisor.)

(a) $__________ ÷ $__________ = ________%

(b) percent increase = ________%

- - - - - - - - - - - - - - -

(a) $570 ÷ $475 = 1.20 = 120%; (b) 20%

49. In the case of a decrease we will have something less than 100% when we divide by the older figure.

Suppose Harris' apple crop dropped from 92,000 tons last year to 68,000 tons this year. What percent decrease does this represent? Since last year's crop decrease, it must be the divisor.

(a) $__________ ÷ $__________ = ________%

(b) percent decrease = 100% - ________% = ________%

(Record your answer to nearest tenth of 1%.)

- - - - - - - - - - - - - - - -

(a) 68,000 ÷ 92,000 = .739 = 73.9%; (b) 100% - 73.9% = 26.1%

50. M. Holman's salary was cut from $175 to $140 a week. What percent decrease does this represent?

- - - - - - - - - - - - - - - -

$140 ÷ $175 = .80 = 80%, 100% - 80% = 20% decrease

51. Example: John's salary was increased 12%. If he earns $890.40 a month now, what was his salary before the increase? What was the amount of increase?

Solution: $890.40 = 112% of John's former salary. His salary before the increase (100%) = $890.40 ÷ 1.12 = $795.00. Therefore, the increase = $890.40 - $795.00 = $95.40.

Suppose John's salary were increased 8%. What was his salary before the increase and what was the amount of the increase? Show the solution here:

- - - - - - - - - - - - - - - -

$890.40 = 108% of his former salary
His salary before the increase = $890.40 ÷ 1.08 = $824.44
Increase = $890.40 - $824.44 = $65.96

52. Receipts for this year at local football games decreased 16% over the year before. What were the receipts last year if this year's receipts amounted to $1629.60? In this case, since there was a decrease, $1629.60 = (100% - 16%) = 84% of last year's receipts.

(a) Then last year's receipts (100%) =

(b) The amount of decrease =

- - - - - - - - - - - - - - - -

(a) Last year's receipts = $1629.60 ÷ .84 = $1940.00
(b) Amount of decrease = $1940 - $1629.60 = $310.40

SELF-TEST

Before you go to the next chapter take this Self-Test. Compare your answers with those given at the end of the test.

1. Change the following percents to decimals and fractions or mixed numbers.

	Decimal	Fraction or Mixed Number
(a) 82%	______	______
(b) $15\frac{1}{2}\%$	______	______
(c) $6\frac{1}{8}\%$	______	______
(d) $208\frac{1}{3}\%$	______	______
(e) 225%	______	______

2. Change the following decimals to percents and fractions or mixed numbers.

	Percent	Fraction or Mixed Number
(a) .06	______	______
(b) .126	______	______
(c) 3.08	______	______
(d) .0025	______	______
(e) .925	______	______

3. Change the following fractions or mixed numbers to decimals and percents.

	Decimal	Percent
(a) $3\frac{1}{2}$	______	______
(b) $\frac{5}{12}$	______	______
(c) $\frac{7}{8}$	______	______
(d) $\frac{5}{80}$	______	______
(e) $\frac{4}{5}$	______	______

4. Record all amounts corrected to 2 decimals (if uneven) and all rates as whole numbers with any remainder shown as a fraction.

   Examples: $22\frac{1}{2}\%$ of 961 = 216.225 Record as 216.23

   15% of $2100 = $1785 Record as $1785.00

   If rate is 33.33...% Record as $33\frac{1}{3}\%$

   (a) Find $12\frac{1}{2}\%$ of 168.

   (b) $\frac{1}{2}$ is what percent of $\frac{5}{6}$?

   (c) $\frac{1}{2}\%$ of 6200 is how much?

   (d) 92 increased by 5% is how much?

   (e) $21\frac{1}{2}$ decreased by 2% = ?

   (f) $37\frac{1}{2}$ is what percent of $12\frac{1}{2}$?

   (g) 479.5 less 20% equals what amount?

   (h) 110% of 36 is how much?

   (i) .08% of 62 = ?

   (j) 59.07 is 22% of what number?

5. (a) Mr. Ames earns $765 per month and saves 7% of this amount. How much does he save in a year?

   (b) The sales for the Baines Market amounted to $25,000 last month and $32,000 this month. What was the percent increase?

   (c) Harvey Brown spent 26% of his salary on rent, which amounted to $180.70. How much did he earn each month?

Answers to Self-Test

Compare your answers to the Self-Test with the answers given here. If you miss any questions, review the frames indicated in parentheses before continuing to the next chapter. If you think you need more practice, additional problems are offered in the Appendix.

1. (a) .82, $\frac{41}{50}$; (b) .155 or $15\frac{1}{2}$, $\frac{31}{200}$; (c) .06125 or $.06\frac{1}{8}$, $\frac{49}{800}$;

   (d) $2.08\frac{1}{3}$, $2\frac{1}{12}$; (e) 2.25, $2\frac{1}{4}$ (Frames 5, 7)

2. (a) 6%, $\frac{3}{50}$; (b) 12.6%, $\frac{63}{500}$; (c) 308%, $3\frac{2}{25}$; (d) .25% or $\frac{1}{4}$%, $\frac{1}{400}$;

   (e) 92.5% or $92\frac{1}{2}$%, $\frac{37}{40}$ (Frames 6, 7)

3. (a) 3.5, 350%; (b) $.41\frac{2}{3}$, $41\frac{2}{3}$%; (c) .875, 87.5% or $87\frac{1}{2}$%;

   (d) .0625, 6.25% or $6\frac{1}{4}$%; (e) .80, 80% (Frame 9)

4. (a) $.125 \times 168 = 21$ (b) $\frac{1}{2} \div \frac{5}{6} = .60 = 60\%$

   (c) $.005 \times 6200 = 31$ (d) $92 \times 1.05 = 96.6$

   (e) $21.5 \times .98 = 21.07$ or $21.5 \times .02 = .43$, $21.5 - .43 = 21.07$

   (f) $37\frac{1}{2} \div 12\frac{1}{2} = 3 = 300\%$

   (g) $479.5 \times \frac{4}{5} = 383.6$ or $479.5 - (\frac{1}{5} \times 479.5) = 479.5 - 95.9 = 383.6$

   (h) $1.10 \times 36 = 39.6$ (i) $.0008 \times 62 = .0496$

   (j) $59.07 \div .22 = 268\frac{11}{22} = 268\frac{1}{2}$ (Frames 17-52)

5. If your answers do not agree with those shown, check your calculations before referring to the solutions that follow.

   (a) \$642.60 (Frame 23); (b) 28% (Frame 27); (c) \$695 (Frame 18)

   Solutions to Question 5
   (a) $765 \times .07 \times 12 = \$642.60$
   (b) $32,000 \div 25,000 = 1.28$, increase = 28% or, since these amounts are even thousands, divide 32 by 25;
   or $32 - 25 = 7$, $7 \div 25 = .28$ or 28%
   (c) $\$180.70 \div .26 = \$695$

CHAPTER FIVE

Solving Word Problems

OBJECTIVES

This chapter will give you some general guidelines for solving word problems and some practice in applying math shortcuts to practical situations. You will learn to use this five-step procedure for solving these problems.

- Identify exactly what answer you must find (Step 1).
- List the known facts and the unknown facts (Step 2).
- Describe in words the relationship between the known facts and the unknown facts (Step 3).
- Change that description from words into numbers and solve the problem using whatever shortcuts apply (Step 4).
- Prove your answer by a different method, if possible (Step 5).

We will follow this procedure, step-by-step, through some examples. Then you can apply these five steps to the practice problems that follow.

EXAMPLES

1. ALWAYS READ A PROBLEM CAREFULLY AT LEAST TWICE. Then you are ready to apply the five-step procedure for solving the problem. Consider the following example.

   If Mr. James deposited $3146.75 in his checking account this month and spent $2036.50 during the same period, how much is the balance approximately?

Step 1: Identify exactly what answer you must find.

What answer must you find?

----

the approximate balance of Mr. James' checking account

Step 2: List the known facts and the unknown facts.

Look again at the example and list what you know and what you do not know.

Known Facts	Unknown Facts

Known Facts: deposited $3146.75, spent $2036.50
Unknown Facts: approximate balance at end of month

Step 3: Describe in words the relationship between the known facts and the unknown facts.

For this problem, we might say: The balance equals the amount deposited less the amount spent. Although this statement is correct, we only need the approximate balance, not the exact balance. So in this case we could say:

The approximate balance equals the approximate amount deposited less the approximate amount spent.

Step 4: Change the description in Step 3 from words into numbers and solve the problem, using whatever shortcuts apply.

$________ - $________ = $________ (approximate balance)

- - - - - - - - - - - - - - -

To nearest $1000 $3000 - $2000 = $1000 (approximate balance)
To nearest $100 $3100 - $2000 = $1100 (approximate balance)

Step 5: Prove your answer by a different method if possible.

To prove an answer in subtraction we know that the answer (difference) plus the subtrahend equals the minuend.

$________ + $________ = $________

- - - - - - - - - - - - - - -

$1000 + $2000 = $3000 (nearest $1000)
$1100 + $2000 = $3100 (nearest $100)

2. ALWAYS READ THE PROBLEM CAREFULLY AT LEAST TWICE. Then apply the procedure outlined in the preceding example to solve it. Try this problem.

Miss Brown typed, on an average, $2\frac{1}{2}$ stencils per hour for 8 hours. If she was paid $3.50 per hour, what was the cost of typing per stencil?

Step 1: Identify exactly what answer you must find.

What answer must you find?

- - - - - - - - - - - - - - -

the cost of typing one stencil

Step 2: List the known facts and the unknown facts.

Look again at the problem and list what you already know and what you don't know.

Known Facts   Unknown Facts

- - - - - - - - - - - - - - -

Known Facts: typed $2\frac{1}{2}$ stencils per hour, typed for 8 hours, paid \$3.50 per hour (you might have listed "the typist's name is Miss Brown" but unfortunately that won't help solve this problem)
Unknown Facts: cost of typing per stencil

Step 3: Describe in words the relationship between the known facts and the unknown facts.

For this problem we might say: The cost of typing each stencil equals the cost per hour (\$3.50) divided by the number of stencils typed per hour ($2\frac{1}{2}$).

Notice, that in describing the relationship in words, we did not use all of the known facts. Many problems include unnecessary information. Part of the trick in solving word problems is to identify what is needed and what is not needed. What known facts in this problem are unnecessary to the solution?

- - - - - - - - - - - - - - -

Miss Brown typed for 8 hours (this fact is unnecessary because you already know how many she typed per hour).

Step 4: Change that description from words into numbers and solve the problem, using whatever shortcuts apply.

Often it is a good idea to estimate your answer to help avoid a large error in calculation. In Step 3, we said: The cost of typing each stencil equals the cost per hour (\$3.50) divided by the number of stencils typed per hour ($2\frac{1}{2}$).

Change that statement into numbers and solve, using shortcuts if they apply.

\$________ ÷ ________ = ________ (cost of typing per stencil)

- - - - - - - - - - - - - - -

$$\$3.50 \div 2\frac{1}{2} = \overset{.70}{\cancel{3}.\cancel{5}\cancel{0}} \times \frac{2}{\cancel{5}} = \$1.40, \text{ or } \$3.50 \times .40 = \$1.40$$

Notice that by using cancelling in one case and the decimal form in the other, you could probably solve this problem in your head.

Step 5: Prove your answer by a different method if possible.

We can prove the answer to the above problem this way: If the cost of typing one stencil is $1.40 and Miss Brown types $2\frac{1}{2}$ stencils in an hour, how much does she earn per hour?

Proof: $1.40 x $2\frac{1}{2}$ = $3.50

3. Try this example. BE SURE TO READ THE PROBLEM CAREFULLY AT LEAST TWICE before proceeding with the solution.

   James earned $510 in commissions during December. His commission rate was 15%. What were his total sales.

   Step 1: Identify exactly what answer you must find.

   What answer must you find?

   - - - - - - - - - - - - - - -

   total sales. We might have said "total sales made by Mr. James" although it is not necessary to add his name since he is the only person involved in this problem.

   Step 2: List the known facts and the unknown facts.

Known Facts	Unknown Facts

   - - - - - - - - - - - - - - -

   Known Facts: commission = $510, commission rate = 15%
   Unknown Facts: total sales

   Step 3: Describe in words the relationship between the known facts and the unknown facts.

   (a) Since this problem deals with percentage, you should immediately think of the three basic elements in any percentage problem. What are they?

   _______________, _______________, _______________

   - - - - - - - - - - - - - - -

   base, rate, percentage

(b) What is the formula for percentage?

- - - - - - - - - - - - - - -

base x rate = percentage

(c) What two elements are known and what element is missing? The rate = 15% but what is the commission? Is it the base or the percentage? We know that total sales will be greater than the commission. Therefore, the commission must equal the percentage since it is part of the total sales. Then, total sales x commission rate = commission (dollars), or

________________ ÷ __________ = total sales.

- - - - - - - - - - - - - - -

commission ÷ rate = total sales

Step 4: Change that description from words into numbers and solve the problem, using whatever shortcuts apply.

_______ ÷ _______ = _______ (total sales)

- - - - - - - - - - - - - - -

\$510 ÷ .15 = \$3400 (total sales)

Step 5: Prove your answer by a different method, if possible.

We know that we can prove a division problem by multiplication.

Therefore, _______ x _______ = _______.

- - - - - - - - - - - - - - -

\$3400 x .15 = \$510 (In multiplying by .15 did you use shortcuts? Take 10% first and then add $\frac{1}{2}$ of it to get 15%. That is, 3400 x .15 = (3400 x .10) + (3400 x .05) = 340 + 170 = 510.)

4. READ THE FOLLOWING PROBLEM CAREFULLY AT LEAST TWICE and then proceed with Step 1.

A house was sold for \$22,528 which was $\frac{3}{8}$ more than it cost the owner. What did the owner pay for the house?

Step 1: Identify exactly what answer you must find.

What answer must you find?

- - - - - - - - - - - - - - -

what the owner paid for the house

Step 2: List the known facts and the unknown facts.

Look again at the problem and list what you know and what you don't know.

Known Facts | Unknown Facts

- - - - - - - - - - - - - - -

Known Facts: selling price = \$22,528, selling price = $\frac{3}{8}$ more than the cost price
Unknown Facts: cost price of home

Step 3: Describe in words the relationship between the known facts and the unknown facts.

Look at the information in Step 2. Try describing this relationship.

- - - - - - - - - - - - - - -

selling price = cost price + $\frac{3}{8}$ of cost price (or cost price = selling price - $\frac{3}{8}$ of cost price)

Step 4: Change the description from words into numbers and solve the problem, using whatever shortcuts apply.

$$\$22{,}528 = \text{cost} + \frac{3}{8} \text{ of cost}$$

We can solve this problem by two methods, either by using eighths or percentage.

(a) Using Eighths: Your problem is to reduce the "cost + $\frac{3}{8}$ of cost" to a common factor--in this case, eighths. How many eights are there in a whole number? The answer is 8. Therefore,

$$\text{"cost} + \frac{3}{8} \text{ of cost"} = \frac{8}{8} + \frac{3}{8} = \frac{11}{8}$$

From this, we know that "cost + $\frac{3}{8}$ cost" = \$22,528. Then, if $\frac{11}{8}$ = \$22,528, what is $\frac{1}{8}$? What is $\frac{8}{8}$ or the total cost?

$$\frac{1}{8} = \$22{,}528 \div 11 = \$2048, \text{ then } \frac{8}{8} \text{ (cost)} = \$2048 \times 8 = \$16{,}384$$

or think of $\frac{11}{8}$ths as 11 eighths = \$22,528
then 1 eighth = \$22,528 ÷ 11 = \$2048
and 8 eighths = \$2048 x 8 = \$16,384

If you do not understand why you divide by 11 (in this case) to obtain $\frac{1}{8}$, forget fractions for the moment. Suppose you bought 3 hats for \$21.00; how much did one hat cost? _______ How did you get the answer? ____________________

- - - - - - - - - - - - - - - -

\$7; divide by 3

Now suppose 11 automobiles cost \$22,528; what does one automobile cost? _______ How did you get the answer? ____________________

- - - - - - - - - - - - - - - -

\$2048; divide \$22,528 by 11

Now go back to the solution (a) shown above. Instead of hats or automobiles, we are talking about eights. Also, 11 eighths is the same as $\frac{11}{8}$.

(b) <u>Using</u> <u>Percentage</u>: Instead of thinking in terms of eighths, you can think of the cost as equal to 100%. Since $\frac{3}{8}$ = $37\frac{1}{2}$%, then cost + $\frac{3}{8}$ of cost = $137\frac{1}{2}$%.

$137\frac{1}{2}$% = \$22,528, then 100% (base) = \$22,528 ÷ 1.375 = \$16,384

(Unless you were using a calculating machine, the first method is easier.)

Step 5: Prove your answer by a different method if possible.

Let us see if the difference between the selling price and the cost price is equal to $\frac{3}{8}$ of the cost price.

selling price $______________

cost price $______________

difference $______________ $= \frac{3}{8}$ of cost price?

- - - - - - - - - - - - - - -

selling price	$22,528	
cost price	16,384	
difference	$ 6,144	Does $6144 = $\frac{3}{8}$ of cost price?

$\frac{3}{8}$ x $16,384 = $6144. Therefore the answer is correct.

PRACTICE PROBLEMS

We will now do some practice problems. See if you can solve them using the five-step procedure. After working each problem, check your answer on page 136. If it does not agree with the one shown, go over your arithmetic and method of solution again. Check your numbers in the solution with the original problem to be sure you copied them correctly. Did you transpose any numbers? If your answer still does not agree with the one given, refer to the worked out solutions on page 137.

1. Deliveries of milk to the Green Haven Cheese Factory from five local dairies were as follows: 2470 gallons, 3240 gallons, 2210 gallons, 1570 gallons, and 3765 gallons. How many gallons, approximately, were delivered to the factory?

   ____

2. What is the average daily pay per employee in a given week for the Ace Box Co. if the total daily payroll for 25 employees during the week equaled the following amounts: $675, $650, $700, $630, and $625?

   ____

3. Mrs. Harvey assembles, on an average, 217 items per day at 12¢ each. How much does she earn in a week if she maintains this average for 6 days?

____

4. The taxes on a piece of property last year were $675.16. This year they amounted to $\frac{7}{5}$ of that figure. How much were the taxes this year?

____

5. Monte Highland District conducted a drive to raise funds for the local orphanage. How much was collected during the first week of the drive if donations were received in the following amounts: $16.50, $2.00, $5.00, $12.50, $0.50, $35.00, $5.00, and $10.00?

____

6. Mrs. Jones' budget allows 12% of her income for entertainment and recreation. If she spent $324 for this purpose last year, what was her income?

____

7. Mr. Whitman had a $\frac{2}{3}$ interest in a manufacturing concern. He sold $\frac{2}{5}$ of his interest for $6000. At the same rate, what was (a) the total amount of his interest and (b) the total worth of the manufacturing concern?

(a) __________

(b) __________

8. Mr. Bradshaw bought 200 used tires for $175 at an auction. He sold $\frac{1}{4}$ of them for $60.00, 16 at $7.50 each, 51 at $10.00 each, and junked the rest. (a) How many tires did he throw away? (b) What was his total profit if the cost of handling this transaction amounted to $65.00?

   (a) ________

   (b) ________

9. It has been estimated that the school population of Anderson County will increase by 25% during the next year. If the present enrollment is 3164, how many students will be enrolled next year?

   ____

10. Mr. Smith receives a 2% bonus on all sales he makes over $300 in any month. If his sales amount to $450 during a certain month, what is the amount of his bonus?

   ____

11. After making contributions to various charities, an organization had $3562.50 left, which represented 95% of its total funds. How much did the organization give to charities?

   ____

12. Mr. Haynes bought a house for $25,500 and made a down payment of 15%. What is the amount of the mortgage (balance due) on the house?

   ____

13. Mr. Thomas spent $\frac{1}{4}$ of his salary for housing, $\frac{1}{3}$ for food, $\frac{1}{6}$ for entertainment, and $\frac{2}{9}$ for clothing. What fractional part did he have left for other needs?

   ____

14. A realtor purchased a house for $16,400. After spending $560 on improvements, he sold it for 120% of the purchase price. What was his profit?

____

15. Mr. Smith owned $\frac{1}{5}$ of a piece of property. If the total value of the property was $19,266, (a) what was his share worth? (b) If he sold $\frac{1}{3}$ of his share for $1500, what was his profit?

(a) __________

(b) __________

16. If a town had a population of 3500 in 1940 and a population of 1200 in 1950, what was the percent of decrease (nearest tenth of 1%)?

____

17. S. J. Cain has 40% of his investments in common stock, 20% in savings and loan accounts, 15% in bonds, and the balance in real estate. The real estate investment amounts to $16,500. What is the total of his investments and how much is invested in stock, savings and loan accounts, and bonds, respectively?

stock __________

savings & loan __________

bonds __________

real estate __________

Total __________

18. A man worked the following number of hours each day during a 5-day week: $6\frac{1}{2}$, $7\frac{3}{4}$, $8\frac{1}{2}$, 8, and $7\frac{2}{3}$, respectively. How many hours did he work and how much did he earn if he was paid at the rate of $6.20 an hour?

hours ________

earned ________

19. Mr. Eaton, Mr. Hayes, and Mr. Huff owned a business jointly and shared profits and losses in proportion to their investments which were $25,000, $15,000, and $30,000, respectively. What fractional part did each man own in the business? If profits amounted to $2682.50 for the current year, how much did each man receive as his share?

	Fraction	Profit
Mr. Eaton	________	________
Mr. Hayes	________	________
Mr. Huff	________	________

20. The sales for R & J Co. were $3500 more this year than last year--an increase of 16%. What were the total sales for last year? For this year?

last year ________

this year ________

21. A profit of $3160 is to be divided among A, B, and C in the ratios of $\frac{1}{2}$, $\frac{1}{3}$, and $\frac{1}{6}$. How much should each receive?

A ________

B ________

C ________

22. Mr. Brown owned $\frac{3}{7}$ of a store. He sold $\frac{1}{3}$ of his share for $13,000. At the same rate, what was the value of the store?

23. The tax on Mr. Mann's home was $500. This represented an increase of $300 over the year before. What was the rate of increase?

24. Mr. Ames spent 22% of his salary on rent, which amounted to $132. How much did he earn each month?

25. The monthly telephone bills for the Banes Optical Co. for 6 months were: $47.16, $59.25, $63.75, $43.50, $59.25, and $87.16. What was the average monthly telephone bill during this time?

26. Mrs. Haynes ordered carpet for her living room. If it cost $6.78 a square yard and her living room contained 125 square yards, what was the approximate cost?

27. Sales for the Rockwell Mfg. Co. amounted to $376,000 last year. During the first 6 months of this year the monthly sales were $37,000, $15,000, $26,500, $37,500, $28,750, and $32,500, respectfully. How much must the sales total for the remainder of the year to equal last year's figure?

28. Visitors to the county museum during the first week after it opened were as follows: Sunday, 869; Monday, 456; Tuesday, 317; Wednesday, 725; Thursday, 294; Friday, 375; Saturday, 961. What was the average daily attendance?

29. Mr. Haverson works in a shoe factory. What are the (a) actual and (b) approximate number of shoes he handles in a week if he inspects 1590 pairs on Monday, 1625 on Tuesday, 1670 on Wednesday, 1650 on Thursday, and 1640 on Friday?

(a) __________

(b) __________

30. The students at Lane High School collected $6102 to send the school band on a trip. (a) If there are 678 students enrolled, what is the approximate amount collected from each student (nearest dollar)? (b) What is the actual average amount collected from each student?

(a) __________

(b) __________

ANSWERS TO PRACTICE PROBLEMS

1. 13,300 or 13,000   2. $26.24   3. $156.24
4. $945.22   5. $86.50   6. $2700
7. (a) $15,000, (b) $22,500   8. (a) 83, (b) $450.00
9. 3955   10. $3.00   11. $187.50
12. $21,675.00   13. 1/36   14. $2720
15. (a) $3853.20, (b) $215.60   16. 65.7%
17.

stock	$26,400
s & l	13,200
bonds	9,900
real estate	16,500
total	$66,000

18. hours: 38 5/12
earned: $238.18

19.

Mr. Eaton	5/14	$ 958.04
Mr. Hayes	3/14	$ 574.82
Mr. Huff	6/14	$1149.64

20. last year: $21,875
this year: $25,375

21. total = $3160; A, $1580.00; B, $1053.33; C, $526.67
22. $91,000   23. 150%   24. $600
25. $60.01   26. $910   27. $198,750
28. 571   29. (a) 16,350, (b) 16,400
30. (a) $8.00, (b) $9.00

SOLUTIONS TO PRACTICE PROBLEMS

Although solutions are shown for all problems, the solutions for the first fifteen problems are arranged in the five-step set-up used in the beginning of this chapter.

1. Step 1: Answer required: approximate number of gallons delivered to the factory

   Step 2: Known facts: gallons delivered by the 5 local dairies: 2470, 3240, 2210, 1570, and 3765
   Unknown facts: approximate total number of gallons delivered

   Step 3: Approximate total gallons delivered equals the total of individual approximate deliveries

   Step 4:

2500		2000
3200		3000
2200	or	2000
1600		2000
3800		4000
13300		13000

   Step 5: Check addition by reverse addition and, if you wish, add the actual amounts and use as a check against the estimates or approximate totals.

2. Step 1: Answer required: Average daily pay per employee

   Step 2: Known facts: 25 employees, daily payroll for each day of week; that is, \$675, \$650, \$700, \$630, and \$625
   Unknown facts: average daily pay per employee

   Step 3: Average daily payroll equals total payroll for the week divided by the number of days and the average pay per employee per day equals the average payroll divided by the number of employees.

   Step 4:
   \$ 675
   650
   700
   630
   625
   \$3280 total weekly payroll

   \$3280 ÷ 5 = \$656 average daily payroll (Did you divide by 10 and multiply by 2?)

   \$656 ÷ 25 = \$26.24 average daily pay per employee (Did you use shortcut--that is, divide by 100 and multiply by 4?)

   We could also solve this problem as follows:

   $$\frac{\overset{656}{\cancel{3280}}}{5 \times 25} = \$26.24$$

Step 5: Proof: average total daily payroll = \$26.24 x 25 = (\$26.24 x 100) ÷ 4 = \$2624 ÷ 4 = \$656
total payroll for the week = \$656 x 5 = (\$656 x 10) ÷ 2 = \$6560 ÷ 2 = \$3280
Notice that shortcuts are used in both operations. (The second step should be done mentally.) Check the total 3280 by reverse addition.

3. Step 1: Answer required: total earnings for the week
Step 2: Known facts: assembles an average of 217 items per day, paid 12¢ each
Unknown facts: total earnings for the week
Step 3: Total weekly earnings equals earnings per day multiplied by number of days.
Step 4: Total earnings for the week = (217 x .12) x 6 = \$156.24
Step 5: Check your multiplication by multiplying 217 by 6 to obtain the total number of items per week (217 x 6 = 1302). If she assembled 1302 items during the week and was paid 12¢ per item, then her earnings for the week = 1302 x .12 = \$156.24.

4. Step 1: Answer required: taxes this year
Step 2: Known facts: taxes last year = \$675.16, taxes this year = $\frac{7}{5}$ of last year's taxes
Unknown facts: Amount of taxes this year
Step 3: Taxes this year equal $\frac{7}{5}$ of last year's taxes.
Step 4: Taxes this year = $\frac{7}{5}$ of \$675.16 = $\frac{7}{5}$ x 135.032 = \$945.224 = \$945.22
Step 5: If this year's taxes = $\frac{7}{5}$ of last year's taxes, this means that they were increased by $\frac{2}{5}$.
Proof: $\frac{2}{5}$ x \$675.15 = \$270.06 = \$270.06
Increase = \$945.22 - \$675.16 = \$270.06

5. Step 1: Answer required: amount collected during the week
Step 2: Known facts: amounts collected each day
Unknown facts: total collected for the week
Step 3: The total collections for the week equals the sum of the daily collections.

Step 4: $16.50
2.00
5.00
12.50
0.50
35.00
5.00
10.00
$86.50 total collected

Step 5: Check your total by reverse addition.

6. Step 1: Answer required: Mrs. Jones' income
   Step 2: Known facts: spent 12% of income for entertainment, entertainment equaled $324; Unknown facts: total income
   Step 3: 12% of the total income equals $324, therefore the total income must equal $324 divided by 12%.
   Step 4: .12 x income = $324, then income = $324 ÷ .12 = $2700
   Step 5: To check this answer, multiply the total income by 12% to see if it equals $324.
   Proof: .12 x $2700 = $324

7. Step 1: Answer required: (a) the total amount of Mr. Whitman's interest in the manufacturing concern and (b) the total worth of the manufacturing concern.
   Step 2: Known facts: Mr. Whitman held $\frac{2}{3}$ interest in company and sold $\frac{2}{5}$ of his interest for $6000
   Unknown facts: (a) total amount of Mr. Whitman's interest
   (b) total worth of mfg. concern
   Step 3: (a) Mr. Whitman's total interest in the company = $\frac{2}{5} + \frac{3}{5}$ where $\frac{2}{5}$ = $6000
   (b) total worth of the manufacturing concern = $\frac{2}{3} + \frac{1}{3}$ (where $\frac{2}{3}$ = Mr. Whitman's total interest)
   Step 4: (a) $\frac{2}{5}$ = $6000, $\frac{1}{5}$ = $6000 ÷ 2 = $3000, $\frac{5}{5}$ or total = 5 x $3000 = $15,000 or $6000 ÷ $\frac{2}{5}$ = $\overset{3000}{\not{6}\not{0}\not{0}\not{0}} \times \frac{5}{\not{2}}$ = $15,000
   (b) $\frac{2}{3}$ = $15,000, $\frac{3}{3}$ (100%) = $15,000 ÷ $\frac{2}{3}$ = $\overset{7,500}{\not{1}\not{5},\not{0}\not{0}\not{0}} \times \frac{3}{\not{2}}$ = $22,500 (Notice use of shortcuts.)

Step 5: Proof: Mr. Whitman had $\frac{2}{3}$ interest in company, therefore $\frac{2}{3}$ x \$22,500 = \$15,000 (interest in company). He sold $\frac{2}{5}$ of interest for \$6000, so $\frac{2}{5}$ x 15,000 = \$6000, which is correct.

8. Step 1: Answer required: number of tires thrown away and the profit

Step 2: Known facts: bought 200 tires for \$175, sold $\frac{1}{4}$ of them for \$60, sold 16 at \$7.50 each, and 51 at \$10.00 each. Cost of transaction \$65.00
Unknown facts: number of tires junked and the profit

Step 3: The number of tires thrown away equals total number purchased less the number sold. Profit equals total income from sales less purchase price and handling cost.

Step 4: (a) Tires junked or thrown away =
$200 - [(\frac{1}{4} \times 200) + 16 + 51)] = 200 - (50 + 16 + 51) =$
$200 - 117 = 83$

(b) $\frac{1}{4}$ x 200 = 50 tires \$ 60.00
16 x \$7.50 = 120.00
51 x \$10.00 = 510.00
amount from sales \$690.00
cost = \$175.00 + \$65.00 = \$240.00
profit = \$690.00 - \$240.00 = \$450.00

Step 5: Proof: You should check the total number of tires sold by reverse addition. Check the subtraction by adding 83 to 117. Does it equal 200? (Did you look at your problem again to verify the number of tires sold and the prices?) Check your multiplication:

$\not{3}\not{0}$
$\not{1}\not{2}\not{0}$ = \$7.50 and 510 ÷ 51 = \$10.00
$\not{1}\not{0}$
$\not{4}$

Check cost: Does \$240 - \$65 = \$175? It does
Check profit: Does \$450 + \$240 = \$690? It does

9. Step 1: Answer required: number of students enrolled next year

Step 2: Known facts: present enrollment 3164, increase of 25% next year; Unknown facts: next year's enrollment

Step 3: The present enrollment (100%, or 3164) plus 25% more equals next year's enrollment.

Step 4: The enrollment next year = 125% of 3164 = 1.25 x 3164 = 3955, <u>or</u>
3164 x .25 = 3164 ÷ 4 = 791 increase next year
3164 + 791 = 3955 total enrollment next year

Step 5: To check this problem, we know that if 3955 = 125% or $\frac{5}{4}$ of this year's enrollment, then

$$3955 \div \frac{5}{4} = \overset{791}{\cancel{3955}} \times \frac{4}{\cancel{5}} = 3164 \text{ present enrollment}$$

10. Step 1: Answer required: amount of bonus
    Step 2: Known facts: 2% bonus on all sales over $300 in any month; sales equaled $450 during one month
    Unknown facts: amount of bonus
    Step 3: The bonus equals 2% of sales made over $300 during the month.
    Step 4: bonus = .02 x ($450 - $300) = .02 x $150 = $3.00
    Step 5: The quickest method of checking this answer is to go over the calculations shown. However, it may also be checked as follows:
    (.02 x $450) - (.02 x $300) = $9.00 - $6.00 = $3.00

11. Step 1: Answer required: amount given to charities
    Step 2: Known facts: $3562.50 left = 95% of total funds
    Unknown facts: amount given to charities
    Step 3: The amount given to charities equals total funds less $3562.50.
    Step 4: We need to find 100% before we can find the amount given to charities. Then, if 95% = $3562.50, 100% = $3562.50 ÷ .95 = $3750.00.* If 95% is left, then 5% was given to charities. Therefore, amount given to charities = $3750 x .05 = $187.50.
    Step 5: The total funds less amount given to charities equals $3562.50.
    Proof: $3750.00 - $3562.50 = $187.50

12. Step 1: Answer required: amount of mortgage or balance due on the house after the down payment had been made
    Step 2: Known facts: purchase price = $25,500, down payment = 15% of purchase price
    Unknown facts: balance due on the house
    Step 3: Balance due equals the purchase price less the down payment.
    Step 4: Balance due = $25,500 - (.15 x $25,500) = $25,500 - $3825 = $21,675
    Step 5: If 15% of the purchase price was the down payment, then 85% of the purchase price was left (balance due).
    Proof: .85 x $25,500 = $21,675

---

*See page 115, Finding the Base or 100%, if you had difficulty here.

13. Step 1: Answer required: fractional part of salary left after housing, food, entertainment, and clothing.

Step 2: Known facts: $\frac{1}{4}$ of salary was spent on housing, $\frac{1}{3}$ of salary was spent on food, $\frac{1}{6}$ of salary was spent on entertainment, $\frac{2}{9}$ of salary was spent on clothing

Unknown facts: fractional part of salary left for other things

Step 3: The part left for other things equals his whole salary less the total spent for housing, food, entertainment, and clothing.

Step 4:
$$\begin{array}{cc} \frac{1}{4} & \frac{9}{36} \\ \frac{1}{3} & \frac{12}{36} \\ \frac{1}{6} & \frac{6}{36} \\ \frac{2}{9} & \frac{8}{36} \\ \hline & \frac{35}{36} \end{array}$$

$$\text{amount left} = \frac{36}{36} - \frac{35}{36} = \frac{1}{36}$$

Step 5: Since we do not have a sum of money to compare to, the best way to check your answer is to go over your addition of fractions.

14. Step 1: Answer required: profit made on sale of house

Step 2: Known facts: Purchase price = \$16,400, improvements = \$560, selling price = 120% of purchase price

Unknown facts: profit on sale

Step 3: Profit equals selling price less (purchase price plus improvements).

Step 4: Selling price = 120% of \$16,400 = 1.20 x \$16,400 = \$19,680

Purchase price + improvements = \$16,400 + \$560 = \$16,960

Profit = \$19,680 - \$16,960 = \$2720

Step 5: Since the selling price = 120% of cost, the profit equals 20% of cost less improvements.

Proof: 20% or $\frac{1}{5}$ of \$16,400 = \$3280 and \$3280 - \$560 = \$2720

15. Step 1: Answer required: (a) how much Mr. Smith's share was worth, (b) his profit when he sold $\frac{1}{3}$ of his share

Step 2: Known facts: owned $\frac{1}{5}$ of a piece of property, total value of property = \$19,266, sold $\frac{1}{3}$ of his share for \$1500

Unknown facts: (a) worth of share, (b) profit on sale

Step 3: (a) The worth of Mr. Smith's share equals $\frac{1}{5}$ of value of property. (b) Profit realized on sale equaled the selling price of $\frac{1}{3}$ of his share less its value.

Step 4: (a) value of share $= \frac{1}{5} \times \$19{,}266 = \$3853.20$

(b) $\frac{1}{3} \times \$3853.20 = \$1284.40$ (value of $\frac{1}{3}$ of his share)

profit = \$1500.00 - \$1284.40 = \$215.60

Step 5: (a) To check your answer to this part of the problem, multiply what Mr. Smith's share is worth by 5 to see if it equals total value of property.

Proof: \$3853.20 x 5 - \$19,266 (Did you use shortcut?)

(b) The profit plus value of $\frac{1}{3}$ of share = selling price.

Proof: \$215.60 + \$1284.40 = \$1500.00

16. Decrease = 3500 - 1200 = 2300, 2300 ÷ 3500 = .657 = 65.7%
Proof: .657 x 3500 = 2299.50 or 2300 corrected to whole numbers

17. 40% + 20% + 15% = 75%, therefore 25% = \$16,500 and 100% = 4 x \$16,500 = \$66,000

stock	40% of \$66,000 = \$26,400
savings & loan	20% of \$66,000 = \$13,200
bonds	15% of \$66,000 = \$ 9,900
real estate	25% of \$66,000 = \$16,500
Total	\$66,000

(check by reverse addition)

18.

$6\frac{1}{2}$	$\frac{6}{12}$
$7\frac{3}{4}$	$\frac{9}{12}$
$8\frac{1}{2}$	$\frac{6}{12}$
8	
$7\frac{2}{3}$	$\frac{8}{12}$
2	
$38\frac{5}{12}$	$\frac{29}{12} = 2\frac{5}{12}$

$38\frac{5}{12} \times \$6.20 = \$238.18$

or $38.4167 \times \$6.20 = \$238.18$

$38\frac{5}{12}$ hours worked (check by reverse addition)

19. $25,000 + $15,000 + $30,000 = $70,000

Mr. Eaton's share = $25,000 ÷ $70,000 = $\frac{5}{14}$

Mr. Hayes' share = $15,000 ÷ $70,000 = $\frac{3}{14}$

Mr. Huff's share = $30,000 ÷ $70,000 = $\frac{6}{14}$

Then,

Mr. Eaton received $\frac{5}{14}$ x $2682.50 = $ 958.04

Mr. Hayes received $\frac{3}{14}$ x $2682.50 = $ 574.82

Mr. Huff received $\frac{6}{14}$ x $2682.50 = <u>$1149.64</u>

Total $2682.50

The fact that the total equals the amount distributed as profit is a check against multiplication and addition. Also, 5 + 3 + 6 = 14 which is a check on the fractional part each man owned.

20. If 16% = $3500, then 100% (last year) = $3500 ÷ .16 = $21,875.
This year = $21,875 + $3500 = $25,375

Proof:	this year	$25,375	
	last year	21,875	
	difference	$ 3,500	(increase of this year over last year)

21. $\frac{1}{2} + \frac{1}{3} + \frac{1}{6}$ = $3160

$\frac{3}{6} + \frac{2}{6} + \frac{1}{6}$ = $3160

A = $\frac{3}{6}$ x $3160 = $1580.00

B = $\frac{2}{6}$ x $3160 = $1053.33

C = $\frac{1}{6}$ x $3160 = <u>$ 526.67</u>

Total $3160.00

Proof: A + B + C = total amount distributed which verifies the multiplication. Also, notice that B's share = two times C's share and A's share = B's share + C's share.

22. Total value of share = $13,000 x 3 = $39,000

If $\frac{3}{7}$ = $39,000, then $\frac{1}{7}$ = $13,000 and 100% or $\frac{7}{7}$ = 7 x $13,000 = $91,000, <u>or</u> 100% could be obtained in one operation as follows:

$$100\% = \$39{,}000 \div \frac{3}{7} = \$\overset{13{,}000}{\cancel{39{,}000}} \times \frac{7}{\cancel{3}} = \$91{,}000$$

Proof: total value of store = $91,000

$\frac{3}{7}$ x $91,000 = $39,000 (value of Mr. Brown's share)

$\frac{1}{3}$ x $39,000 = $13,000 (selling price of $\frac{1}{3}$ of Mr. Brown's share)

23. $500 - $300 = $200 tax last year
$300 ÷ $200 = 1.50 or 150% increase rate
Proof: 150% of $200 = $300 increase
$200 + $300 = $500 tax this year

24. 22% = $132, therefore 100% = $132 ÷ .22 = $600 (salary)
Proof: .22 x $600 = $132.00 (rent)

25. $ 47.16
59.25
63.75
43.50
59.25
87.16
$360.07 total of bills for 6 months

$360.06 ÷ 6 = $60.01 (average monthly bill)

Proof: Check total by reverse addition.

26. 7 x $130 = $910

Proof: $910 ÷ 7 = $130

27. $ 37,000
15,000
26,500
37,500
28,750
32,500
$177,250 total sales for first 6 months

last year $376,000
6 months 177,250
$198,750 sales needed for rest of year to meet last year's figure

Proof: Check total of 6-month's sales by reverse addition. Check subtraction by adding the answer to 6-month's figure. It should equal last year's sales.

28. 869
456
317
725
294
375
961
3997 total visitors for the week

3997 ÷ 7 = 571 average daily attendance

Proof: 571 x 7 = 3997 Check attendance total by reverse addition.

29.
```
     1590
     1625
     1670
     1650
     1640
(a)  8175  pairs handled in a week
     8175 x 2 = 16,350  actual number of shoes handled in a week

     1600          Multiply by 2 since there are two shoes per pair
     1600          and we want to know how many shoes, not pairs,
     1700          that were handled in the week.
     1700
     1600
(b)  8200 x 2 = 16,400  approximate number of shoes handled in
                        a week
```

Proof: Check totals by reverse addition.

30. (a) $6000 ÷ 700 = $8   (b) $6102 ÷ 678 = $9

Proof: (a) 700 x 8 = $5600 (approximate amount collected)

(b) 678 x 9 = $6102 (exact amount collected)

Using shortcuts, multiply 678 by 10 and subtract 678 as follows:

```
678 x 10 = 6780
678 x  1 =  678
           6102
```

APPENDIX

Practice Problems

ADDITION BY GROUPING 10S

See how many minutes it takes you to complete these problems. Add each one, grouping by 10s.

```
1.   823        2. 4528        3. 1074
     569            980           4208
    5460           5990            878
    5421            659            582
    ----           ----           ----

4. .0078        5. 87.87       6.  18.67
   .6954            9.19          102.21
   .0708           41.19           55.55
   .1538             .89            3.78
   -----          -----          ------

7. 42.68        8. 27,889      9.  31.27
    5.33            1,342           8.99
    2.47           31,178          45.44
   77.12            8,556          82.14
   -----           36,230           9.56
                   ------         118.06
                                   67.55
                                   63.16
                                  ------

10. 879.25
     23.30
    638.17
     61.53
    452.00
     78.56
    ------
```

- - - - - - - - - - - - - - - -

Answers

```
1.     8 2 3
       5 6 9
     5 4 6 0
     5 4 2 1
    12 2 7 3

2.   4 5 2 8
       9 8 0
     5 9 9 0
       6 5 9
    12 1 5 7

3.   1 0 7 4
     4 2 0 8
       8 7 8
       5 8 2
     6 7 4 2

4.   .0 0 7 8
     .6 9 5 4
     .0 7 0 8
     .1 5 3 8
     .9 2 7 8

5.    8 7. 8 7
        9. 1 9
      4 1. 1 9
         . 8 9
     13 9. 1 4

6.    1 8. 6 7
    1 0 2. 2 1
      5 5. 5 5
        3. 7 8
    1 8 0. 2 1

7.   4 2. 6 8
        5. 3 3
        2. 4 7
      7 7. 1 2
     12 7. 6 0

8.   2 7 8 8 9
       1 3 4 2
     3 1 1 7 8
       8 5 5 6
     3 6 2 3 0
    10 5,1 9 5

9.     3 1. 2 7
          8. 9 9
        4 5. 4 4
        8 2. 1 4
          9. 5 6
      1 1 8. 0 6
        6 7. 5 5
        6 3. 1 6
      4 2 6. 1 7

10.  8 7 9. 2 5
       2 3. 3 0
     6 3 8. 1 7
       6 1. 5 3
     4 5 2. 0 0
       7 8. 5 6
    21 3 2. 8 1
```

If you completed these problems in 3 or 4 minutes, you are doing fine and, if all of your answers are correct, good for you. If you missed only one, that is good, too. Since the number of problems completed correctly per minute is important, if you completed fewer than 7 correctly in 3 minutes, you need to practice.

SUBTOTALS

Record the subtotal where indicated and total. Check your answers by adding the column without the use of subtotals.

1.

211		
89		
664	______	subtotal
8117		
988		
1052	______	subtotal
782		
3198	______	subtotal
		total

2.

306		
121		
96	______	subtotal
1022		
821	______	subtotal
56		
727		
8021		
102	______	subtotal
		total

- - - - - - - - - - - - - - -

1.

211		
89		
664	964	subtotal
8117		
988		
1052	10157	subtotal
782		
3198	3980	subtotal
15101	15101	total

2.

306		
121		
96	523	subtotal
1022		
821	1843	subtotal
56		
727		
8021		
102	8906	subtotal
11272	11272	total

CROSSFOOTING

Complete the following report and check by adding each column down and each line across. Check the grand total by adding the total column and total line. The sum should be the same.

	A	B	C	D	Total
Line A	323	566	821	900	______
B	56	821	919	1066	______
C	567	789	507	899	______
Total	______	______	______	______	______ Grand total

- - - - - - - - - - - - - - - -

Line totals		
	A	2610
	B	2862
	C	2762
Grand total		8234

Column totals		
	A	946
	B	2176
	C	2247
	D	2865
Grand total		8234

ESTIMATING ADDITION AND SUBTRACTION

1. Find the exact and estimated total to the nearest $100 for the following.

   Exact: $326 + $115 + $25 + $720 + $196 = ________

   Estimate: ______ + ______ + ______ + ______ + ______ = ________

2. What is the estimate to the nearest $100 of the following list by grouping two or more numbers?

Problem:		Estimate:
	3168	________
	420	________
	9006	________
	35	________
	826	________
	1207	________
	99	________
	14761	

3. The approximate difference between $2662.43 and $834.42 is $________.

- - - - - - - - - - - - - - - -

1. Exact, $1382; estimate, $300 + $100 + $700 + $200 = $1300

2. 3200
   400
   9000
   900
   1300
   14800

3. $2662.43 - $834.42 = $1828.01
   estimate is $2700 - $800 = $1900

MULTIPLICATION

Work the following problems as quickly as you can, using shortcuts when appropriate. Record answers in the spaces provided. Time yourself. Check your answers with those shown at the end of this exercise.

1. 23 x 25 =	________	2. 319 x 5 =	________
3. 27.5 x 101 =	________	4. .0675 x 10 =	________
5. 976 x 50 =	________	6. 51.32 x 1000 =	________
7. 700 x 125 =	________	8. 3.167 x 100 =	________
9. 74 x 110 =	________	10. 90 x 96 =	________
11. 46 x 35 =	________	12. 38 x 62 =	________
13. 47.36 x 80.2 =	________	14. .189 x 14.6 =	________
15. 37.5 x 4.06 =	________		

- - - - - - - - - - - - - - - -

Answers

1. 575	2. 1595	3. 2777.5	4. .675	5. 48,800
6. 51,320	7. 87,500	8. 316.7	9. 8140	10. 8640
11. 1610	12. 2356	13. 3798.272	14. 2.7594	15. 152.250

Solutions

1. 23 x 25 = (23 x 100) ÷ 4 = 2300 ÷ 4 = 575
2. 319 x 5 = (319 x 10) ÷ 2 = 3190 ÷ 2 = 1595
3. 27.5 x 101 = (27.5 x 100) + (27.5) = 2750 + 27.5 = 2777.5
4. .0675 x 10 = .675
5. 976 x 50 = (976 x 100) ÷ 2 = 97,600 ÷ 2 = 48,800
6. 51.32 x 1000 = 51,320
7. 700 x 125 = (700 x 1000) ÷ 8 = 700,000 ÷ 8 = 87,500
8. 3.167 x 100 = 316.7
9. 74 x 110 = (74 x 100) + (74 x 10) = 7400 ÷ 740 = 8140
10. 90 x 96 = (96 x 100) - (96 x 10) = 9600 - 960 = 8640
11. 46 x 35 = 1610
12. 38 x 62 = 2356
13. 47.36 x 80.2 = 3798.272
14. .189 x 14.6 = 2.7594
15. 37.5 x 4.06 = 152.250 or $\frac{3}{8}$ x (4.06 x 100) = $\frac{3}{8}$ x 406 = 152.25

ESTIMATING MULTIPLICATION AND DIVISION

1. When estimating the answer to 356 x 8135, if 356 is rounded to 400, then 8135 is rounded to ________. The estimate is __________; the exact answer is __________.

2. What is your estimate for the following?

   (a) 120.6 x 8.15 __________

   (b) 916 x 15.9 __________

3. In estimating the answer to the following problem, if the divisor is rounded to the nearest 10s, then the dividend must be rounded to the nearest _______.

   The estimate for 360 ÷ 19 is ____________________.

4. When estimating the answer for division of decimals, treat the numbers as _______________ numbers.

   What is the (a) exact, to nearest tenth, and (b) estimated answer to 56.8 ÷ 8.5?

   (a) ____

   (b) ____

- - - - - - - - - - - - - - - -

1. 8100; estimate is 3,240,000; exact answer is 2,896,060
2. (a) 121 x 8 = 120 x 8 = 960; (b) 916 x 16 = 920 x 20 = 18400
3. 100s; 400 ÷ 20 = 20
4. whole numbers; (a) 56.8 ÷ 8.5 = 6.7; (b) 57 ÷ 9 = 60 ÷ 9 = $6\frac{2}{3}$

DIVISION

1. Complete the following:

   (a) 132 ÷ 10 = (b) 2563 ÷ 1000 =

   (c) 6.03 ÷ 100 = (d) 36.08 ÷ 100 =

   (e) .365 ÷ 10 = (f) 25.63 ÷ 1000 =

   (g) 164 ÷ 125 = (h) 83 ÷ 5 =

(i) $8.60 \div 25 =$ (j) $36.6 \div 16\frac{2}{3} =$

(k) $45 \div 11\frac{1}{9} =$ (l) $1450 \div 50 =$

2. Find the average for each of the following series.

(a) 83, 56, 90, 100, 265, 45, 55, 260

(b) 2.56, 13.4, .026, .45, .008

Answers

1. (a) 13.2 (b) 2.563 (c) .0603 (d) .3608
   (e) .0365 (f) .02563 (g) 1.312 (h) 16.6
   (i) .344 (j) 2.196 (k) 4.05 (l) 29.0
2. (a) $119\frac{1}{4}$ (b) 3.2888

Solutions

1. (a) & (e) To divide by 10, move decimal one place to left.
   (c) & (d) To divide by 100, move decimal two places to left.
   (b) & (f) To divide by 1000, move decimal three places to left.
   (g) $164 \div 125 = (164 \div 1000) \times 8 = .164 \times 8 = 1.312$
   (h) $83 \div 5 = (83 \div 10) \times 2 = 8.3 \times 2 = 16.6$
   (i) $8.60 \div 25 = (8.60 \div 100) \times 4 = .086 \times 4 = .344$
   (j) $36.6 \div 16\frac{2}{3} = (36.6 \div 100) \times 6 = .366 \times 6 = 2.196$
   (k) $45 \div 11\frac{1}{9} = (45 \div 100) \times 9 = .45 \times 9 = 4.05$
   or $.45 \times 9 = (.45 \times 10) - (.45 \times 1) = 4.50 - .45 = 4.05$
   (l) $1450 \div 50 = (1450 \div 100) \times 2 = 14.5 \times 2 = 29.0$
2. (a) $83 + 56 + 90 + 100 + 265 + 45 + 55 + 260 = 954$, $954 \div 8 = 119\frac{1}{4}$
   (b) $2.56 + 13.4 + .026 + .45 + .008 = 16.444$, $16.444 \div 5 = 3.2888$

3. In these problems, find the quotients for the following, correct to 2 decimals. Division must be carried to 1 more decimal than is required in the answer. If the third decimal is less than 5, drop it; if it is more than 5, increase the second decimal by 1: for example, (a) $17.428 \div .24 = 72.616$ (to 3 decimals) or 72.62 corrected to 2 decimals; (b) $5.6 \div 92 = .060$. Record as .06.

(a) $.24 \div 2.7 =$ (b) $72 \div 26.4 =$

(c) $8.7 \div 15.7 =$ (d) $6.6 \div .75 =$

(e) $23.86 \div 375 =$ (f) $148 \div .026 =$

4. Record all answers correct to the nearest thousandth. Carry quotient to 4 decimals and increase or drop the third decimal as illustrated. (1) 48   58 = .8275. Record as .828. (2) 6.795   12 = .5662. Record as .566.

   (a) .90 ÷ 3.2 = (b) 100 ÷ 3.85 =

   (c) 48 ÷ .16 = (d) 62 ÷ 1.50 =

   (e) .648 ÷ 4.86 = (f) 49 ÷ .54 =

- - - - - - - - - - - - - - - -

Answers

3. (a) .09; (b) 2.73; (c) .55; (d) 8.80; (e) .06; (f) 5692.31
4. (a) .281; (b) 25.974; (c) 300.000; (d) 41.333; (e) .133; (f) 90.741

Solutions

3. (corrected to second decimal)

(a)
```
         .0 8 = .09
  2.7 | .2 4 0
         2 1 6
           2 4
```

(b)
```
            2.7 2 = 2.73
  2 6.4 | 7 2.0 0 0
          5 2 8
          1 9 2 0
          1 8 4 8
              7 2 0
              5 2 8
              1 9 2
```

(c)
```
             .5 5 = .55
  1 5.7 | 8.7 0 0
          7 8 5
            8 5 0
            7 8 5
              6 5
```

(d)
```
           8.8 0 = 8.80
  .7 5 | 6.6 0 0
         6 0 0
           6 0 0
           6 0 0
```

or $6.6 \div \frac{3}{4} = \overset{2.2}{\not 6.\not 6} \times \frac{4}{3} = 8.80$

(e)
```
              .0 6 = .06
  3 7 5 | 2 3.8 6
          2 2 5 0
            1 3 6
```

(f)
```
             5 6 9 2.3 0 = 5629.31
  .0 2 6 | 1 4 8.0 0 0 0 0
           1 3 0
             1 8 0
             1 5 6
               2 4 0
               2 3 4
                   6 0
                   5 2
                     8 0
                     7 8
                     2 0
```

4. (answers corrected to nearest thousandth or third decimal)

(a)
```
          .2 8 1 = .281
3.2 |.9 0 0 0
      6 4
      2 6 0
      2 5 6
          4 0
          3 2
            8
```

(b)
```
               2 5.9 7 4 = 25.974
3.8 5 |1 0 0.0 0 0 0 0
          7 7 0
          2 3 0 0
          1 9 2 5
            3 7 5 0
            3 4 6 5
              2 8 5 0
              2 6 9 5
                1 5 5 0
                1 4 4 0
                  1 1 0
```

(c)
```
         3 0 0.0 0 0 = 300.000
.1 6 |4 8.0 0 0 0
```

(d)
```
             4 1.3 3 = 41.333
1.5 0 |6 2.0 0 0
        6 0 0
          2 0 0
          1 5 0
            5 0 0
            4 5 0
              5 0 0
```

(e)
```
               .1 3 3 = .133
4.8 6 |.6 4 8 0 0
         4 8 6
         1 6 2 0
         1 4 5 8
           1 6 2 0
```

(f)
```
            9 0.7 4 0 = 90.741
.5 4 |4 9.0 0 0 0 0
       4 8 6
         4 0 0
         3 7 8
           2 2 0
           2 1 6
               4 0
```

FRACTIONS

1. Add the following:

(a)
$$16\frac{1}{8}$$
$$2\frac{5}{8}$$
$$196\frac{3}{8}$$
$$21$$

(b)
$$21\frac{1}{6}$$
$$8\frac{2}{3}$$
$$19\frac{5}{8}$$
$$6\frac{1}{4}$$

2. Subtract the following:

(a) $196\frac{2}{3} - 29$

(b) $85\frac{7}{8} - 30\frac{3}{8}$

(c) $16\frac{4}{9} - 8\frac{5}{9}$

(d) $473\frac{2}{3} - 209\frac{5}{7}$

3. Multiply:

(a) $9\frac{1}{3} \times 6\frac{1}{7} =$

(b) $8\frac{5}{6} \times 4\frac{6}{15} =$

(c) $12\frac{2}{5} \times 3\frac{3}{4} =$

(d) $8\frac{1}{2} \times 6\frac{2}{3} =$

4. Divide:

(a) $6\frac{2}{3} \div 2\frac{1}{12} =$

(b) $15\frac{1}{2} \div 2\frac{1}{2} =$

(c) $10\frac{2}{4} \div 3\frac{1}{2} =$

(d) $8 \div 4\frac{1}{4} =$

(e) $\frac{3\frac{1}{8}}{4\frac{1}{6}} =$

(f) $\frac{2\frac{1}{7}}{6\frac{2}{8}} =$

Answers

1. (a) $236\frac{1}{8}$; (b) $55\frac{17}{24}$
2. (a) $167\frac{2}{3}$; (b) $55\frac{1}{2}$; (c) $7\frac{8}{9}$; (d) $263\frac{20}{21}$
3. (a) $57\frac{1}{3}$; (b) $38\frac{13}{15}$; (c) $46\frac{1}{2}$; (d) $56\frac{2}{3}$
4. (a) $3\frac{1}{5}$; (b) $6\frac{1}{5}$; (c) 3; (d) $1\frac{15}{17}$; (e) $\frac{3}{4}$; (f) $\frac{12}{35}$

Solutions

1. (a)

$$\begin{array}{r} 16\frac{1}{8} \\ 2\frac{5}{8} \\ 196\frac{3}{8} \\ 21 \\ \hline 236\frac{1}{8} \end{array}$$

$$\frac{1}{8} + \frac{5}{8} + \frac{3}{8} = \frac{9}{8} = 1\frac{1}{8}$$

(b)

$$\begin{array}{r|l} 21\frac{1}{6} & \frac{4}{24} \\ 8\frac{2}{3} & \frac{16}{24} \\ 19\frac{5}{8} & \frac{15}{24} \\ 6\frac{1}{4} & \frac{6}{24} \\ \hline 55\frac{17}{24} & \frac{41}{24} = 1\frac{17}{24} \end{array}$$

$$\begin{array}{r|rrrr} 2 & 6 & 3 & 8 & 4 \\ \hline 2 & 3 & 3 & 4 & 2 \\ \hline 3 & 3 & 3 & 2 & 1 \\ \hline & 1 & 1 & 2 & 1 \end{array}$$

LCD = 2 x 2 x 3 x 2 = 24
(This could have been found by examination.)

2. (a)

$$\begin{array}{r} 196\frac{2}{3} \\ -\ 29 \\ \hline 167\frac{2}{3} \end{array}$$

(b)

$$\begin{array}{r} 85\frac{7}{8} \\ -\ 30\frac{3}{8} \\ \hline 55\frac{4}{8} = 55\frac{1}{2} \end{array}$$

(c)

$$\begin{array}{rr} 16\frac{4}{9} = & 15\frac{13}{9} \\ -\ 8\frac{5}{9} & 8\frac{5}{9} \\ \hline & 7\frac{8}{9} \end{array}$$

(d)

$$\begin{array}{r|l} 47\overset{2}{\cancel{3}}\frac{2}{3} & \frac{\overset{35}{\cancel{14}}}{21} \\ -\ 209\frac{5}{7} & \frac{15}{21} \\ \hline 263\frac{20}{21} & \frac{20}{21} \end{array}$$

3. (a) $9\frac{1}{3} \times 6\frac{1}{7} = \frac{\overset{4}{\cancel{28}}}{3} \times \frac{43}{7} = \frac{172}{3} = 57\frac{1}{3}$

(b) $8\frac{5}{6} \times 4\frac{6}{15} = \frac{53}{\cancel{6}} \times \frac{\overset{11}{\cancel{66}}}{15} = \frac{583}{15} = 38\frac{13}{15}$

(c) $12\frac{2}{5} \times 3\frac{3}{4} = \frac{\overset{31}{\cancel{62}}}{\cancel{5}} \times \frac{\overset{3}{\cancel{15}}}{\underset{2}{\cancel{4}}} = \frac{93}{2} = 46\frac{1}{2}$

(d) $8\frac{1}{2} \times 6\frac{2}{3} = \frac{17}{\cancel{2}} \times \frac{\overset{10}{\cancel{20}}}{3} = \frac{170}{3} = 56\frac{2}{3}$

4\. (a) $6\frac{2}{3} \div 2\frac{1}{12} = \frac{20}{3} \div \frac{25}{12} = \frac{\overset{4}{\cancel{20}}}{\cancel{3}} \times \frac{\overset{4}{\cancel{12}}}{\underset{5}{\cancel{25}}} = \frac{16}{5} = 3\frac{1}{5}$

(b) $15\frac{1}{2} \div 2\frac{1}{2} = \frac{31}{2} \div \frac{5}{2} = \frac{31}{\cancel{2}} \times \frac{\cancel{2}}{5} = \frac{31}{5} = 6\frac{1}{5}$

(c) $10\frac{2}{4} \div 3\frac{1}{2} = \frac{42}{4} \div \frac{7}{2} = \frac{\overset{21}{\cancel{42}}}{\underset{2}{\cancel{4}}} \times \frac{\cancel{2}}{7} = \frac{21}{7} = 3$

(d) $8 \div 4\frac{1}{4} = 8 \div \frac{17}{4} = 8 \times \frac{4}{17} = \frac{32}{17} = 1\frac{15}{17}$

(e) $3\frac{1}{8} \div 4\frac{1}{6} = \frac{25}{8} \div \frac{25}{6} = \frac{\cancel{25}}{\underset{4}{\cancel{8}}} \times \frac{\overset{3}{\cancel{6}}}{\cancel{25}} = \frac{3}{4}$

(f) $2\frac{1}{7} \div 6\frac{2}{8} = \frac{15}{7} \div \frac{50}{8} = \frac{\overset{3}{\cancel{15}}}{7} \times \frac{\overset{4}{\cancel{8}}}{\underset{\underset{5}{\cancel{25}}}{\cancel{50}}} = \frac{12}{35}$

FRACTIONS

5\. Reduce the following fractions to their lowest terms.

(a) $\frac{36}{124} =$ (b) $\frac{55}{120} =$

(c) $\frac{9}{135} =$ (d) $\frac{21}{560} =$

6\. Change the following improper fractions to whole or mixed numbers.

(a) $\frac{28}{9} =$ (b) $\frac{83}{10} =$

(c) $\frac{50}{8} =$ (d) $\frac{107}{5} =$

7\. Change the following mixed numbers to improper fractions.

(a) $19\frac{2}{3} =$ (b) $5\frac{1}{6}$

(c) $21\frac{7}{9} =$ (d) $10\frac{1}{8} =$

8. Change the following fractions to the higher terms indicated.

(a) $\frac{2}{3} = \frac{\quad}{15}$ (b) $\frac{5}{6} = \frac{\quad}{24}$

(c) $\frac{8}{9} = \frac{\quad}{108}$ (d) $\frac{3}{4} = \frac{\quad}{12}$

9. Change the following fractions or mixed numbers to decimals (carry answers to three decimal places).

(a) $6\frac{1}{9} =$ (b) $12\frac{3}{8} =$

(c) $\frac{5}{12} =$ (d) $\frac{23}{16} =$

10. Change the following decimals to fractions.

(a) $.41\frac{2}{3} =$ (b) $.16\frac{2}{3} =$

(c) $.44\frac{4}{9} =$ (d) $.045 =$

Answers

5. (a) $\frac{9}{31}$; (b) $\frac{11}{24}$; (c) $\frac{1}{15}$; (d) $\frac{3}{80}$

6. (a) $3\frac{1}{9}$; (b) $8\frac{3}{10}$; (c) $6\frac{1}{4}$; (d) $21\frac{2}{5}$

7. (a) $\frac{59}{3}$; (b) $\frac{31}{6}$; (c) $\frac{196}{9}$; (d) $\frac{81}{8}$

8. (a) $\frac{10}{15}$; (b) $\frac{20}{24}$; (c) $\frac{96}{108}$; (d) $\frac{9}{12}$

9. (a) 6.111; (b) 12.375; (c) .417; (d) $1\frac{7}{16} = 1.4375 = 1.438$

10. (a) $.41\frac{2}{3} = \frac{41\frac{2}{3}}{100} = \frac{\overset{5}{\cancel{125}}}{3} \times \frac{1}{\underset{4}{\cancel{100}}} = \frac{5}{12}$

(b) $.16\frac{2}{3} = \frac{16\frac{2}{3}}{100} = \frac{\cancel{50}}{3} \times \frac{1}{\underset{2}{\cancel{100}}} = \frac{1}{6}$

(c) $.44\frac{4}{9} = \frac{44\frac{4}{9}}{100} = \frac{\overset{4}{\cancel{400}}}{9} \times \frac{1}{100} = \frac{4}{9}$

(d) $.045 = \frac{45}{1000} = \frac{9}{200}$

PERCENTAGE

Changing Percents to the Decimal Form, Mixed Number, or Fraction, and Vice Versa

1. Change each of the following numbers to the percent form:

(a) $.25 = 25\%$ (b) $.33\frac{1}{3} =$ (c) $1.65 =$

(d) $.05 =$ (e) $.005 =$ (f) $3 =$

(g) $1.06\frac{1}{2} =$ (h) $1.25 =$ (i) $3\frac{1}{2} =$

(j) $.16 =$ (k) $2.05 =$ (l) $.008 =$

(m) $10 =$ (n) $.0036 =$ (o) $1\frac{1}{4} =$

(p) $.90 =$ (q) $.00\frac{1}{2} =$ (r) $.7 =$

(s) $.02\frac{1}{2} =$ (t) $1\frac{1}{2} =$

2. Change each of the following percents to the decimal form:

(a) $2\% = .02$ (b) $200\% =$ (c) $.5\% =$

(d) $\frac{1}{4}\% =$ (e) $1.6\% =$ (f) $1.5\% =$

(g) $16\% =$ (h) $25\% =$ (i) $300\% =$

(j) $16\frac{1}{2}\% =$ (k) $.05\% =$ (l) $32\frac{1}{2}\% =$

(m) $9\% =$ (n) $4\frac{1}{2}\% =$ (o) $568\% =$

(p) $\frac{1}{2}\% =$ (q) $23.2\% =$ (r) $.56\% =$

(s) $1.7\% =$ (t) $115\% =$

3. Change each of the following percents to a fraction or a mixed number:

(a) $25\% = \frac{1}{4}$ (b) $16\frac{2}{3}\% =$ (c) $12\frac{1}{2}\% =$

(d) $11\frac{1}{9}\% =$ (e) $108\frac{1}{3}\% =$ (f) $20\% =$

(g) $5\% =$ (h) $3\frac{1}{2}\% =$ (i) $30\% =$

(j) 125% = (k) 15% = (l) $83\frac{1}{3}\%$ =

(m) 60% = (n) 17% = (o) .22% =

(p) 6% = (q) $\frac{2}{5}\%$ = (r) 38% =

(s) 46% = (t) 375% =

4. Change each of the following fractions or mixed numbers to the percent (%) form.

(a) $\frac{3}{8}$ = 37.5% (b) $\frac{5}{6}$ = (c) $\frac{1}{12}$ =

(d) $\frac{3}{9}$ = (e) $\frac{2}{5}$ = (f) $\frac{5}{12}$ =

(g) $\frac{1}{30}$ = (h) $\frac{1}{25}$ = (i) $\frac{1}{2}$ =

(j) $\frac{3}{4}$ = (k) $1\frac{7}{8}$ = (l) $3\frac{3}{4}$ =

(m) $\frac{1}{6}$ = (n) $\frac{2}{3}$ = (o) $\frac{2}{16}$ =

(p) $\frac{4}{32}$ = (q) $1\frac{2}{5}$ = (r) $\frac{5}{9}$ =

(s) $1\frac{1}{3}$ = (t) $20\frac{1}{2}$ =

- - - - - - - - - - - - - - - - -

Answers

1. (a) 25%; (b) $33\frac{1}{3}\%$; (c) 165%; (d) 5%; (e) .5 or $\frac{1}{2}\%$; (f) 300%; (g) 106.5%; (h) 125%; (i) 350%; (j) 16%; (k) 205%; (l) .8%; (m) 1000%; (n) .36%; (o) 125%; (p) 90%; (q) $\frac{1}{2}\%$; (r) 70%; (s) $2\frac{1}{2}\%$; (t) 150%

2. (a) .02; (b) 2.00; (c) .005; (d) .0025; (e) .016; (f) .015; (g) .16; (h) .25; (i) 3.00; (j) .165; (k) .0005; (l) .325; (m) .09; (n) .045; (o) 5.68; (p) .005; (q) .232; (r) .0056; (s) .017; (t) 1.15

3. (a) $\frac{1}{4}$; (b) $\frac{1}{6}$; (c) $\frac{1}{8}$; (d) $\frac{1}{9}$; (e) $1\frac{1}{12}$; (f) $\frac{1}{5}$; (g) $\frac{1}{20}$; (h) $\frac{7}{200}$; (i) $\frac{3}{10}$; (j) $1\frac{1}{4}$; (k) $\frac{3}{20}$; (l) $\frac{5}{6}$; (m) $\frac{3}{5}$; (n) $\frac{17}{100}$; (o) $\frac{11}{5000}$; (p) $\frac{3}{50}$; (q) $\frac{1}{250}$; (r) $\frac{19}{50}$; (s) $\frac{23}{50}$; (t) $3\frac{3}{4}$

4. (a) $37\frac{1}{2}\%$; (b) $83\frac{1}{3}\%$; (c) $8\frac{1}{3}\%$; (d) $33\frac{1}{3}\%$; (e) 40%; (f) $41\frac{2}{3}\%$; (g) $3\frac{1}{3}\%$; (h) 4%; (i) 50%; (j) 75%; (k) $187\frac{1}{2}\%$; (l) 375%; (m) $16\frac{2}{3}\%$; (n) $66\frac{2}{3}\%$; (o) $12\frac{1}{2}\%$; (p) $12\frac{1}{2}\%$; (q) 140%; (r) $55\frac{5}{9}\%$; (s) $133\frac{1}{3}\%$; (t) 2050%

PERCENTAGE

Finding Base, Rate, and Percentage
(Base x Rate = Percentage)

1. Find the percentage, given the rate and base.
   (a) $16\frac{1}{2}\%$ of \$350.00
   (b) $\frac{1}{2}\%$ of 9260
   (c) 115% of \$240
   (d) 200% of 82
   (e) .02% of 456
2. Find the rate, given the percentage and base. Record whole number with any remainder shown as a fraction.
   (a) 32 is what percent of 96?
   (b) What percent of 64 is 1.92?
   (c) 300 is what percent of 60?
   (d) What percent of 64 is 56?
   (e) What percent of 5000 is 2.5?
   (f) $7.87\frac{1}{2}$ is what percent of 63?
3. Find the base (100%), given the rate and percentage. Record complete number with any remainder shown as a fraction.
   (a) 70 is 25% of what number?

(b) $16\frac{2}{3}$% of \$________ = \$264.

(c) If 272 = 34%, what is 100%?

(d) \$10.00 is $\frac{1}{2}$% of what amount?

(e) \$39 is 75% of what amount?

(f) 125 is 110% of what number?

- - - - - - - - - - - - - - - -

Answers

1. (a) \$57.75; (b) 46.30; (c) \$276.00; (d) 164; (e) .0912
2. (a) $33\frac{1}{3}$%; (b) 3%; (c) 500%; (d) $87\frac{1}{2}$%; (e) .05%; (f) $12\frac{1}{2}$%
3. (a) 280; (b) \$1584; (c) 800; (d) \$2000; (e) \$52; (f) $113\frac{7}{11}$

Index